Thomas W. Wieting
Reed College, 1969.

Studies in the Foundations
Methodology and Philosophy of Science
Volume 1

Delaware Seminar in the Foundations of Physics

Edited by

Mario Bunge

With 5 Figures

Springer-Verlag New York Inc. 1967

William L. Reese

General Editor of Delaware Seminars in the Philosophy of Science,
University of Delaware, Newark, Delaware / USA

Mario Bunge

McGill University
Montreal/Canada

Library of Congress Catalog Card Number 67—16650
Printed in Germany

Title-No. 6760

Introduction to the Delaware Seminars

At its inception the purpose of the Delaware Seminar in the Philosophy of Science was declared to be one of bridging "the growing chasm between the two intellectual cultures of our time." The task of reconciling the scientific and humanistic communities is clearly an important one; and, indeed, may be the appropriate long-range goal of the seminar as it continues to develop through the years. In its first two years, however, the seminar was addressed to the somewhat easier task of exploring differences of opinion among scientists interested in the philosophic implications of scientific subject matter; and differences of opinion among philosophers who had carefully considered various scientific subjects.

Volume III represents a further stage of development. After the first two years it became evident that the area of investigation needed to be specified more precisely; as a result the task of the seminar has become that of foundation study. Correspondingly, the confrontation has changed in character, featuring scientists whose interests do not exclude philosophy and philosophers who are also scientists. Volume III of the Delaware Seminar — the present volume — contains the record of this confrontation with respect to the Foundations of Physics.

William L. Reese

Preface

This volume collects the lectures on the foundations of physics given by eleven scientists at the University of Delaware. It is neither an anthology of disconnected items nor a smoothly running textbook but rather a progress report on a neglected yet vital area of basic physical research, namely foundations research.

The investigation into the foundations of any branch of science is neither loose speculation nor popular science: it is an aspect of scientific research — in fact the deepest-searching part of basic research. Consequently it must be carried out by the scientist himself. Thus whether the time concept is a numerical variable or a function, whether particle mechanics is the primary mechanical theory or a particular case of continuum mechanics, whether transformation formulas are laws of nature or links between equivalent descriptions, whether thermodynamics has been fully reduced to statistical mechanics, whether the field concept is dispensable, whether the covariance principle is a law or a regulative principle, whether quantum mechanics is completely detached from classical physics or contains fragments of it, whether it has annihilated the physical object or given a more complex picture of it, and to which extent are the field variables measurable — all these are technical questions demanding a careful analysis of pieces of recent basic research.

Yet all of these problems and indeed all questions in foundational research are philosophical as well as scientific. They are not philosophical in the popular sense of being cloudy or incapable of being settled by rational argument: they are philosophical in a technical sense. In fact every one of them calls for a logical, a semantical and a methodological analysis and may even involve metaphysical considerations. Thus take the problem of determining the components of the angular momentum in quantum mechanics. According to the current theory they do not commute and this noncommutativity means that it is not possible to make precise measurements of all three components at the same time. Is this interpretation unavoidable or are there alternative interpretations (models) of the same mathematical formulas? This is a semantical problem. In either case, must we say that we cannot measure the angular momentum with precision or rather that the particle has no fully determined angular momentum? Either question is philosophical: in the first case the limits of our knowledge are in question, in the second the existence of properties symbolized by physical variables. Of course it may be that these philosophical questions can only be answered, or perhaps bypassed, by further scientific research concerning, say, the search for an adequate choice of the position and velocity operators or even the complete abandonment of such a search. In any case the questions themselves have a definite philosophical tinge and consequently the research into them calls for a philosophical spirit.

Where shall we turn for a bit of such philosophical spirit: to the pure philosopher? It is a sad yet universally recognized truth that most philosophers are so distant from science that they can hardly be expected to make constructive remarks on it. It is less generally admitted but equally true that most scientists are so strongly and so unwittingly committed to an underdeveloped philosophical credo that they make no effort to improve on it. In fact most of us harbor naive and often anachronistic and even mutually inconsistent philosophical tenets, such as that reality is a dangerous metaphysical monster, that every concept must be defined, that concepts are defined by reference to empirical operations, that laws are rules or prescriptions, that theories are induced from or revealed by data, that they refer only to what can be done in the laboratory and should therefore involve experimentally accessible variables alone, that science describes but does not explain, and so on.

True enough, just as in the cases of religion and politics, the credo is observed by a minority and unawarely violated by the majority. Otherwise there would be no scientific progress. Yet just as in the case of dogmas of other kinds, the Credo of the Innocent Physicist is apt to make itself felt at the crossroads, when some light is needed to see beyond what is known: to spot new problems, to design and try new research lines, and to interpret the results. It may well be that the present obscurities in elementary particle physics and elsewhere are due not only to the inherent difficulties of the subjects and to their novelty but also because an inadequate pathfinder is employed to illumine the scene: namely the Credo of the Innocent Physicist. This tool cannot help gestate the new ideas needed at every point of crisis because it distrusts ideas and trusts only observation — as if scientific observations could be designed and interpreted without theories. If we care for the much-needed stimulation of the emission of new ideas we shall do well to avail ourselves of a suitable philosophical apparatus, one emphasizing that science works with ideas, and indeed that it goes farthest with the more abstract and daring ideas — without forgetting that in the case of physics most such ideas happen to refer to physical objects and are therefore testable by handling such objects whether in a direct manner or not. We need, in sum, a *lasei* — short for Light Amplification by Stimulated Emission of Ideas.

Philosophers may not be able to supply this apparatus but physicists must use part of their work to build it. To begin with we must learn the elements of mathematical logic, a fruit of the marriage of philosophy and mathematics. Both philosophers and mathematicians have built it up and have used it to cleanse their own conceptual equipment. With the help of mathematical logic, a number of mathematical and philosophical muddles have been cleared away and a few but decisive contributions have been made to the theory of theories, to the theory of reference, to the methodology of science (e.g., the criticism of inductivism and operationalism), and to some epistemological and metaphysical problems posed by scientific theories and procedures. Although these general results are important for the foundations of physics, very little of relevance to the specific problems of this discipline has so far resulted from that revolution in philosophy. The reason seems to be that most of the philosophical problems raised by physics are technical and consequently most philosophers shy away from them while physicists do not regard them as "hard" enough to plunge into them.

How about physicists: are they better equipped? Of course they are as far as the raw material is concerned. But this raw material — the body of available physical knowledge — must be handled with certain tools that are not supplied by the conventional training physicists receive. In fact the standard physics curriculum does not include courses on mathematical logic, semantics, metatheory, epistemology, methodology, or even the foundations of physics. Moreover, most physicists are unaware that such tools could be of help in their own scientific work: because traditional philosophy has not been too helpful, they tend to think that every philosophy must be scientifically barren or even nonsensical. What is more, most find it easier to abuse philosophy, both prescientific and scientific, than to become acquainted with it. Yet we all know that arrogance is a bad counselor: that it prevents one from learning and from realizing that something is being done in other fields as well.

By adopting an arrogance traditionally attributed to philosophers, the physicist deprives himself of the light that some good philosophizing might cast on his own research. This arrogance is certainly not the attitude prevailing among mathematicians or even in the advanced quarters of young disciplines like psychology and sociology, where people are taking logical, semantical and methodological questions seriously. Is it not time we took a humbler attitude and a more realistic one, acknowledging that other disciplines are catching up with physics and that we should learn at least as much philosophy as the mathematician, the learning theorist or the mathematical sociologist? Or shall we continue to handle foundation problems in an amateurish way?

Of course not everyone need be concerned mainly with foundation problems. In order to compute or measure one need not normally ask fundamental questions or try to bring logical order to a theory or perform a semantical analysis of it — just as the analyst need not worry about puzzles in set theory or in number theory. But if one does ask fundamental questions — such as whether theories can be built without jumping beyond what can be observed, or whether space and time are dispensable, or whether time reversal is anything more than velocity inversion — then one had better face these problems with the proper equipment. And, whether or not one is interested in getting a deep understanding, whether or not one cares for more than increasing the amount of information, one should be tolerant of scientific endeavors other than one's own.

This plea for intellectual tolerance is less anachronistic than it may sound. In fact there is a tendency nowadays to regard with suspicion any activity which involves neither evaluating integrals nor measuring cross sections. Any effort to get a better understanding of what has been done or to analyze it critically sounds to most ears as parasitic and at best fruitless entertainment rather than as research. One does not normally apply for a grant in order to find out, say, how a given theory should be interpreted. The criterion for the selection of research problems and for the patronage of research into them seems to be that of addition to the available body of knowledge. Questions regarding the analysis and logical organization of scientific knowledge would seem to be regarded at best as fit for vacation periods or for retirement. It seems to have been forgotten that no breakthrough has come out of the blue, that every new idea is preceded by a critical analysis and evaluation of the prevailing ideas; that Galilei was not a laboratory technician but a scholar who spent much of his time criticizing physical theories and discussing foundation problems. It is claimed that nowadays the pace of research is much too hectic for scholars and that these are out of place in science. What is true is that the contract for short-term research does not favor and is not intended to favor any adventure in ideas. Luckily ARCHIMEDES, GALILEI, HUYGENS, LEIBNIZ, NEWTON, EULER, CAUCHY, FARADAY, MAXWELL, BOLTZMANN, EINSTEIN, BOHR, HEISENBERG, SCHRÖDINGER, FERMI, and the other giants who built physics did not depend on such contracts and could afford delving in deep problems.

For the sake of the maturation of science — for its growth in depth and cogency and not only in coverage — let us pay more attention to basic research and let us be more tolerant with regard to the investigation into that very research — namely foundations research. After all, everything in both pure and applied science depends ultimately upon basic research. And basic research can be helped by foundations research at least to the extent of freeing it from unjustifiable strictures stemming from earlier philosophies, still held by most physicists although they have been surrendered by many philosophers.

The editor of this volume bears the responsibility for the preceding lines: no one else should be punished for advocating a scholarly attitude towards scientific research. The reader will find that every contributor to this volume of readings in the foundations of physics has his own ideas on the matter and that no effort has been made to give them a uniform, i.e. dull, appearance. This lack of unanimity concerning a number of fundamental questions — many of them supposedly elementary — is hoped to suggest that the foundations of physics are both much alive and in need of further contributions.

MARIO BUNGE

Contents

1 PETER G. BERGMANN
Foundations Research in Physics 1

2 MARIO BUNGE
The Structure and Content of a Physical Theory 15

3 WALTER NOLL
Space-Time Structures in Classical Mechanics 28

4 CLIFFORD A. TRUESDELL
Foundations of Continuum Mechanics 35

5 HAROLD GRAD
Levels of Description in Statistical Mechanics and Thermodynamics . . 49

6 EDWIN T. JAYNES
Foundations of Probability Theory and Statistical Mechanics 77

7 E. J. POST
General Covariance in Electromagnetism 102

8 PETER HAVAS
Foundation Problems in General Relativity 124

9 RALPH SCHILLER
Relations of Quantum to Classical Physics 149

10 HENRY MARGENAU and JAMES L. PARK
Objectivity in Quantum Mechanics 161

11 PAUL BERNAYS
Scope and Limits of Axiomatics 188

Biographical Notes on the Contributors 192

Chapter 1

Foundations Research in Physics[1]

Peter G. Bergmann

Department of Physics, Syracuse University
Syracuse, New York

Permit me to begin with a sort of sociological point, and this is the lack of communication between professional scientists and professional philosophers. As you all undoubtedly know, philosophy has prided itself for centuries for being "the queen of the sciences". Unfortunately, there has been a long period during which this particular queen has shared the fate of her human colleagues; it has reigned rather than governed. Only, in recent times something worse has happened to her: University communities have begun to question the need for her "civil list", that is her emoluments. Today, humanities departments are frequently poorly supported, and their chief usefulness is often that they provide a number of "gut" courses, in which undergraduates can acquire a number of credit points required for "distribution" without having to memorize foreign vocabularies or being required to exert themselves otherwise. This is, of course, a situation which has hurt the former "subjects" also, though perhaps not in the pocketbook, but certainly in terms of spiritual impoverishment.

I do not think that I need to belabor the fact that the serious scientist frequently encounters philosophical problems in the course of apparently purely "technical" research, that is to say research that is considered strictly within his professional jurisdiction. Among the exact sciences physics is concerned perhaps more than chemistry with far-out problems, and thus we are brought up to face philosophical problems more often than other scientists; though I believe that the biosciences encounter foundations problems, too, perhaps of different varieties than we do. Be that as it may, I have grown up in an environment in which the chairs of theoretical physics and of natural philosophy were in close proximity to each other, both geographically and in spirit, their respective

[1] During the preparation of this lecture and manuscript the author's work was supported in part by the Air Force Office of Scientific Research and by Aerospace Research Laboratories.

holders being the Professors PHILIPP FRANK and RUDOLPH CARNAP, and I know that it is not constitutionally impossible for physicists and philosophers to talk to each other in a language that is mutually comprehensible.

Examples of complete misunderstanding are, unfortunately, all too frequent, and I don't need to quote horrible examples from personal experience. It behooves all of us, not in a posture of condescension but out of sheer necessity for the well-being of our respective disciplines, to rebuild bridges that may have collapsed simply because of long-neglected maintenance. As a physicist, I shall make no attempt to adopt the protective coloration of philosophers' terminology, but sketch, in the language that physicists employ, some of the philosophical problems that we encounter.

1. Epistemology

Let me begin with some of the recurring problems of the relationship between the universe of objects and the subject — observer and experimenter. Whether there "exists" a universe of objects, and whether we, the subjects, are capable of gathering valid information about that universe has been a problem as old as philosophy. Undoubtedly, some aspects of that question are semantic. We must imbue the term "existence" with some meaning that raises the problem above the level of a tautology, and this has been done by various schools in many different ways. The schools of "physicalism" in Europe, and of "operationalism" in this country have tended to take the view that "existence" is indeed void of any significance and that the issues of existence as well as knowability are "meaningless". For some time this point of view was quite popular among scientists. Instead of asking whether the outside universe exists and whether we can ever perceive it, let us proceed with the business of investigating specific scientific problems, to which valid answers can be obtained by the universally accepted procedures known as "the scientific method".

Unfortunately, physicalism does not provide us with unambiguous answers to such questions as the validation of scientific (or everyday) observations. The ravings of the lunatic are *in principle* indistinguishable from the findings of a serious scientist; we are reduced to such methods of distinction as popularity votes. Nor does physicalism have much to contribute to the unraveling of what *is* from the changes brought about in the universe by our experimentation. Finally, MACH's principle of "Denkökonomie" appears insufficient, in a very practical sense, to provide us with a sound foundation from which to analyze what we mean by an *understanding* of observed facts, that is to say what we mean by the formation of scientific hypotheses and theories.

A few examples are in order. MACH, the founder of European physicalism, strenuously opposed the hypothesis of the molecular structure of matter, on the grounds that molecules were not observable (at the time he was right in that), and hence not legitimate building blocks for a physical theory. In other words, molecules were "metaphysical". The young EINSTEIN, who at that time was a great admirer of MACH'S but who was also very much concerned with the molecular explanation of Brownian motion and similar phenomena, ventured to journey to MACH'S Institute to raise this question: Suppose that the theory of molecular structure did no more than predict a relationship between the coefficients of heat conductivity and of viscosity which could be confirmed experimentally, would MACH then accept the molecular hypothesis as a legitimate physical theory? MACH answered affirmatively but apparently considered this contingency so improbable that he continued to oppose the molecular theory in public for some time to come.

Another example from this side of the Atlantic: BRIDGMAN, the father of operationalism, remained opposed to general relativity, on the grounds that in that theory none of the basic elements are directly accessible to "operational" analysis. BRIDGMAN was right, but the same criticism applies to any other physical theory if you look at its building blocks singly.

From these two examples you might think that the only proper conclusion is that philosophers should not meddle with physics, even if they happen to earn a living as professors of physics. I think that such a conclusion would be entirely inappropriate, particularly in view of the development of quantum theory. In my two previous examples, the molecular theory of matter, and BRIDGMAN'S overnaive critique of general relativity, I have merely illustrated that a doctrinaire approach to novel physical theories may blind us to their merits. But both of these instances are concerned with "classical" theories. They are classical in that they are concerned with the universe of objects as something to be explored but given *ab initio*. The effect of the observer's actions on the objects of our physical theory is considered unimportant. — As usual, what is unimportant is not analyzed, or even mentioned; the implicit assumption of negligible reaction of the *ego* on the *id* forms part of our *a priori* collective frame of mind. — Not so in quantum theory. When we make an observation, we affect, irreversibly, the quantum state of the physical system, we "collapse the wave packet". Well, what are we doing in making an observation? A then young colleague of mine, not quite twenty years ago, answered that question as follows: "How to make a quantum observation? Why, you diagonalize the operator, that's obvious". I suppose that is as "operational" an answer as you can hope for, but it is hardly responsive to the question of what it is that we do to a physical system when we make an observation. Is it the quantum state, the Hilbert space vector, that represents physical reality, or is it the actual physical variables, locations and momenta as we know them from classical mechanics? If the latter, is the uncertainty relation of Heisenberg part of the physical reality, or does it represent merely a bound for the completeness of our possible information about that reality? If, on the other hand, the quantum state is to be the sum total of physical reality, how come that our manipulation changes it in a manner that cannot be represented by any possible Schrödinger or Heisenberg equation of motion?

These questions have led to profound discomfort on the part of all physicists who have stood at the cradle of our present theory, with the possible exception of HEISENBERG. NIELS BOHR, ALBERT EINSTEIN, and ERWIN SCHRÖDINGER all have written extensively on the epistemological issues raised by quantum theory. DIRAC, when he was recently asked about his own point of view, replied that he considered quantum theory so ephemeral, so badly in need of a replacement by a more definitive theory, that he felt that it simply was not worth the bother of a profound epistemological critique.

I have no new points of view to contribute, except, perhaps, to say that I believe that the quantum state is most likely not to be taken as part of physical reality, that is as a property of the physical system independent of the extent and quality of our information about it. As a matter of historical accuracy, I should perhaps warn against the expression "the Copenhagen school". As far as I am aware, BOHR himself considered the epistemological questions arising out of quantum theory far from settled; probably the several prominent scientists who at one time or another stayed at his Institute formed varying opinions about these matters, and none of them ought to be assumed to speak for a monolithic consensus.

2. Logical Analysis

According to the textbooks, a proper mathematical theory begins with a statement of all the axioms of the theory, which represent logically arbitrary inventions of the mathematicians, subject only to the single requirement that they be mutually consistent. The remainder of the theory then consists of the deductive derivation of all the theorems that professional competence and ingenuity may be capable of extracting from the assumed axioms. A logically satisfactory physical theory is to be one that permits a similar formulation.

As for mathematical theories, we know today that the requirement of logical consistency may be undecidable, at least in the case of non-trivial structures, and furthermore that the validity of certain proposed conclusions may also be undecidable. Besides, systems of axioms may be devoid of mathematical interest though not inconsistent. A good mathematician will be able to tell you whether a certain theory is "deep", and whether a theorem is "powerful". Thus the business of mathematics requires more than logical neatness, and logic alone is demonstrably incapable of deciding what would appear to be the basic preliminary criteria of validity. Nevertheless, we do not question the import of mathematical research.

When we come to the analysis of physical theories, we are all aware of the historical instances in which the analysis and critique of certain notions has been instrumental for theoretical progress. EINSTEIN'S critique of the concept of simultaneity has been instrumental in resolving the crisis in electrodynamics that had been brought about by the crucial experiments on "ether wind" in the closing years of the nineteenth century. SCHRÖDINGER established that his wave mechanics and HEISENBERG'S matrix mechanics were equivalent in that they were bound to lead to identical predictions of the outcomes of any conceivable experiment. Nevertheless, I believe that the examples I gave earlier demonstrate that not every element of a physical theory that cannot be reduced to observable entities is an idealistic fiction, to be isolated by rigorous axiomatic analysis and to be cast out without mercy. Rather, I think that logical analysis of physical theories can serve the following purposes: (1) We may discover the implicit assumptions within a theoretical structure, in addition to the explicit elements that are obvious to everyone; and (2) we may discover which elements of the theory are essential in its relationship to observable facts of nature and which ones may be discarded without detracting from the theory's physical content.

Elements that are logically superfluous, or redundant, in one theory may have great heuristic value in the development of the theory in new directions. For instance, classical analytic mechanics, with all of its new concepts, such as the notion of an action integral, of canonically conjugate variables, of the Hamiltonian and of the Hamilton-Jacobi function, and with its elaborate transformation theory, does not add one iota to the physical content of mechanics as originally conceived by NEWTON. But I need not demonstrate to an audience of physicists how enormously important all of these concepts have been in the gradual evolution of quantum mechanics as we know it today. As a matter of fact, I believe that quantum mechanics cannot be taught to an intelligent person as a physical theory unless it has been preceded by a pretty sophisticated course in classical analytic mechanics.

In other words, I believe that logical analysis and axiomatic formulation of physical theories are important and useful procedures in that they help us to understand a theory better, in that they highlight possible internal inconsistencies, and in that they may provide hints for further developments. But to rule out the whole of a physical theory on the grounds that it suffers from internal inconsistencies, or on the much weaker charge that it contains elements not susceptible to experimental verification appears to me unjustified.

As a matter of fact, the structure of physical theories appears to me to call for considerably more analysis than mathematical theories. A physical theory contains a number of formal elements that resemble those of a mathematical theory without physical implications, but it has additional aspects which have no analogs in mathematics. Let us take an example that is reasonably well analyzed, say geometrical optics. We have as formal elements "light rays", "transmitting media", a "three-dimensional space", an "index of refraction". The "law of nature" that combines these elements into a formal theory is FERMAT's principle, or alternatively, HUYGENS's principle (the eikonal equation). These two principles may be shown to be equivalent, in that they both lead to systems of differential equations that single out certain curves in the space previously mentioned as possible "ray trajectories". In this form geometric optics is nothing but a very special class of stationary curves in functional analysis. Considering particular congruences of such curves we may, for instance, introduce the concepts of foci and of caustics, but except for our terminology and notation we have introduced no logical elements that deviate in any way from those in use throughout functional analysis.

Obviously, no matter how we elaborate this mathematical structure, we shall not gain within it any elements that provide a connection with observable phenomena, or with the reality that we call the physical universe. To accomplish such a connection we must, first of all, describe roughly what class of physical processes we propose to analyze in terms of our structure. We can do this in the language of the theoretical physicist, who speaks of electrodynamics in general, then of electrodynamics away from electric charges, and who gradually works his way toward the "short wave limit", in which the propagation of electromagnetic radiation is no longer sensitive to the coherence properties of the radiation involved. Making fleeting reference to the WBK method and thus assuring the audience that he is at home in quantum mechanics, he finally settles down to HAMILTON'S optical-mechanical analogy and proceeds to show that an electron microscope ought to be considered formally analogous to an ordinary optical instrument ,though in one case the trajectories are "light rays" in the other "corpuscles".

If we proceed more historically, we might start out with the obvious fact that there are physical phenomena that come to our attention primarily through one of our senses, through the eyes, and that early experience teaches us that light has a tendency to propagate along straight lines, as evidenced by the formation of shadows. Further observation and experimentation teaches us that there are exceptions to rectilinear propagation, that is to say the reflection and the refraction of light. We may delineate the domain of geometric optics further by stating that those phenomena that depend in detail on the nature of the radiation observed, polarization, diffraction, etc., are not to be included in our domain of interest.

Clearly, we must construct a relationship between some of the notions in our formal theory with physical processes; our terminology, such as calling our special curves "light rays", indicates the manner in which such relationships are to be understood. As we become acquainted with physical processes through our senses, or at a later stage through more or less elaborately instrumented procedures, we must establish through what sort of observations we can test the predictions of the formal theory. Typically, we do so through a number of intermediate steps. For instance, we might say that we ascertain the trajectories of light rays by observing the deposition of energy on absorbers intercepting the rays. It may turn out that actually we record the readings of galvanometers registering the output of bolometers; or that we develop, and subsequently analyze with a densitometer, photographic emulsions. That we consider several methods of observation as basically equivalent may indicate our reliance on other physical theories which are not an integral part of geometrical optics. Thus it may turn out that none of the predictions of geometrical optics is susceptible to "direct" testing, that is to measurements independent of other theoretical structures.

Some of these structures may be quite primitive. We measure the location of a point on the ray trajectory with the help of an optical bench along with other accessories, whose use implies our faith in the existence of rigid bodies. This faith is initially not based on a theory of the solid state, but a very precise experiment must take into account thermal expansion and deformation under mechanical stress. This endless criss-crossing of theoretical relationships and empirical knowledge that underlies the "testing" of predictions of even fairly simple theories is typical. In fact, the more fundamental the theory to be tested, the more difficult it becomes to state what one means by the "empirical content" of the theory. Because we are incapable of considering simultaneously all the ramifications of an experiment, we confine ourselves to a domain that is cut off, more or less arbitrarily, from adjoining territory. Intuitively, we believe, we know what we mean by the measurement of any quantity occurring in a physical theory, but I submit that such intuitive knowledge must always be considered preliminary, and subject to amendment.

Permit me to give you an example from my personal experience. In the early days of general relativity the theory was formulated entirely in terms of the metric tensor field defined on the four-dimensional space-time manifold. It was, of course, realized that two formally distinct metric fields might go over into each other under a coordinate transformation, and that in this sense many formally different fields were intrinsically equivalent. Nevertheless, it was thought generally that in any given coordinate system the metric tensor could be measured directly, with the help of (idealized) scales and clocks. In the course of my quest for a quantum theory of gravitation (a quest whose end is not in sight) I gradually came to the realization that the metric tensor is not measurable within the context of general relativity itself. The problem is not the unavailability of appropriate scales and clocks, because these can be approximated to any degree of accuracy. The problem is the identification of world points between which space and time intervals might be determined. To identify a world point it is not sufficient to assign to it four coordinate values (three space and one time coordinate); after all, according to the principle of relativity coordinates have no physical significance. No, we must identify a world point in terms of physical identifying marks, such as the values of other physical fields. In astronomy we identify world points in terms of their relationships to stars. But the stars, being the sources of the gravitational field, cannot serve to identify anything but their own trajectories. And we cannot populate the universe by clouds of idealized observers, or the like, because real observers would have to be represented by material structures (such as instruments, mirrors, etc.), which would modify the gravitational field, and idealized observers without physical reality will hardly serve as building blocks for a realistic theory. In the end we realized that world points might be identified in terms of local curvature; thus the metric tensor might be replaced by the gradient fields of the curvature. We have called such involved structures observables.

At this stage, then, we consider that we have related a mathematical object, the metric field, to something that can in principle be observed if we have succeeded in describing the gravitational field in terms of observables. In this description we have eliminated all references to an arbitrary coordinate system, retaining only elements that would remain unchanged if we replaced the metric field in one coordinate system by the equivalent metric field expressed in terms of a different coordinate system. Thus, in terms of observables, two formally different fields will be in fact inequivalent. Does that mean that we have succeeded in reducing the theory of gravitation to its "guts" elements, that we have obtained a language that tells us directly how to measure any observable by a specific experimental procedure? Far from it. We believe that we have overcome an essential road block in our efforts to separate redundant from essential formal elements in the theory of gravitation, incidentally at a terrific sacrifice in mathematical simplicity. But though I believe that the construction of experimental procedures for the determination of the values of specified observables has been reduced to a technical question, I may eventually turn out to be wrong. We have, as it were, peeled one layer off the logical onion, and our eyes are beginning to tear. We do not know much about the layers that are left.

3. Space and Time

According to KANT, space and time are given to us *a priori*, that is to say prior to any empirical contact with the physical universe. We cannot imagine any experience that is not cast in the space-time framework.

NEWTON thought otherwise. He was well aware of the fact that the properties of space and time were to be ascertained in part, at least, by physical experiments, and he discussed such experiments in justifying what today we should call the Galilean principle of relativity. But not only NEWTON had been aware of the fact that the properties of physical space-time were subject to experimentation. In the course of the nineteenth century mathematicians established that Euclid's geometry was but one of many geometries that were thinkable. The time is long past in which the physicist can prove his mathematical sophistication by pointing out that Euclidean geometry is a special case of Riemannian geometry. The fact of the matter is that quite aside from general relativity physicists have been using non-Riemannian spaces for quite some time, though not as models of physical space-time. One example that comes to mind is phase space. Phase space is non-metric. There is no way of defining the distance between two points in phase space in a manner that will stand up under canonical transformations. What is more, such a definition would serve no useful purpose in any application of the phase space concept. But the phase space still has structure. First of all, the phase space has a topology; we can, and do, conceive of neighborhoods, which will retain their identities under canonical transformations. Moreover, phase space has an invariant volume, and its existence is the formal content of LIOUVILLE's theorem, on which statistical mechanics is based.

But let me return to models of physical space-time. According to the general theory of relativity all reasonably continuous and smooth coordinate systems will serve equally well, not only to identify world points but to formulate the laws of nature. This principle is usually referred to as the principle of covariance. The space that serves as the model of physical space-time in general relativity not only possesses a topology, it is a metric space, albeit with an indefinite metric; — squared distances between world points may turn out to be either positive or negative, time-like or space-like. And the distance between two world points that can be connected by a light signal vanishes. Thus the distance concept that occurs in relativity lacks some of the attributes that we intuitively associate with distance elsewhere. But, to quote "Oklahoma": "Have we gone as far as one can go" in generalizing geometric concepts? And more important, have we gone as far as it is likely we should go in order to have a good mathematical model of physical space-time?

No one knows the answer, but I rather suspect that it should be "no". The topological properties of our current "best" model certainly cannot be tested in any sense that we can specify. What is more, because of the existence of elementary particles and because of the possibility of creating particles in any scattering experiment, we cannot expect that we can resolve without limit all distances by the employment of sufficiently short-wave-length radiations. There exists a universal distance which is defined irrespective of the properties of any particular species of elementary particle. This distance is the Compton wavelength of a particle of such mass that the Compton wavelength equals its gravitational radius. The length is of the order of 10^{-33} cm, and the corresponding mass of the order of 10^{-5} g. For a particle of smaller mass, the Compton wavelength is correspondingly greater, for a particle of larger mass, the gravitational radius would exceed 10^{-33} cm. Thus, there would seem to exist a lower bound for any straight-forward resolution of very small distances.

Conceivably, there exists then a lower bound for the particularization of points and of their geometric relationship to each other. Obligingly, mathematics stands ready to serve the physicist in quest of space models with incomplete resolvability. One class of models that I have become acquainted with are called lattices. Instead of points lattices have as their structural units "patches". These patches can be endowed with properties that make them behave like ordinary point sets if they are of sufficient size, but very differently when they are small. But whether physical theory will ever be driven into the adoption of a lattice model for physical space-time I certainly do not pretend to know. The logical possibilities appear so numerous that one should have considerable guidance from the needs of a physical theory to account for phenomena occurring in the very small in order not to drown in an *embarras de richesse.*

Before I leave the topic of space and time, I should like to discuss briefly the current status of their symmetries. As you know, parity was dethroned in 1957 with the discovery that neutrinos have a definite helicity, which for a given species of neutrinos cannot be reversed. Within recent months experimental results on kaon decay have been published which would seem to indicate that the laws of nature may not be CP-invariant, either. That is to say, the coupling of helicity reversal with charge conjugation (a transition that leads from particles to anti-particles, and *vice versa*) will not leave the laws of particle interaction unchanged. Because it has been shown that in a very wide class of field theories the combination of CP reversal with time reversal (T) must leave the form of the laws unchanged, there is a suspicion that if these experimental results should be confirmed, and if every alternative explanation fails, we shall finally have uncovered an elementary law which will permit us to distinguish intrinsically the future time direction from the past. This issue is so fundamental that we might well await the outcome of further investigations before accepting the ultimate consequences of this discovery as established fact.

An entirely different aspect of the "arrow of time" is presented by thermodynamics and statistical mechanics. The Second Law of thermodynamics asserts that a large number of macroscopic processes can take place only in one direction. These processes are said to be *irreversible*. BOLTZMANN, MAXWELL, GIBBS, and EINSTEIN proposed to explain irreversibility in terms of the statistical nature of macroscopic processes. Whereas BOLTZMANN originally conceived of the randomness of large-scale processes in terms of the large number of constituent particles involved, GIBBS realized that the validity of thermodynamic concepts could not be made dependent merely on the large number of molecules in a given amount of matter; he proposed to extend statistical concepts to physical systems with arbitrarily small numbers of degrees of freedom, provided only that our information on the state of the system (its "phase") was sufficiently incomplete. As his model for incomplete information he introduced the concept of *ensemble*, that is to say a collection of replicas of a given physical system, all of which move in accordance with the laws of mechanics but starting from different initial configurations and momenta. The average values of all physical variables, averaged over the ensemble, were then to tend in the course of time toward values corresponding to thermodynamic equilibrium.

It is not very difficult to show that the combination of the reversible laws of mechanics with Gibbsian statistics does not lead to irreversibility but that the notion of irreversibility must be added as an extra ingredient; in fact its addition calls for very delicate handling if the resulting structure is not to be internally self-contradictory. Many of the proofs of irreversibility to be found in the literature are self-serving. They are either downright false, or they can be used to prove increase of the entropy in both directions, the future and the past, assuming identical conditions at the starting time t_0. Considering further that irreversibility is not observed on ensembles but on individual systems, and that the Gibbsian ensemble is a purely formal construct, unrelated to any physical object, the explanation of irreversibility in nature is to my mind still open. I consider rather promising a proposal by BONDI and GOLD, to the effect that irreversibility is ultimately related to the expansion of our universe, that is to a cosmological circumstance.

4. Cosmology

So far I have talked of the laws of nature as if it were to be understood *a priori* that such laws exist, and our task as scientists is to discover them. I presume that we all understand laws of nature to be regularities that enable us to assert with confidence that physical processes can be repeated if we take sufficient care to reproduce exactly all pertinent conditions.

If we are to believe present-day quantum mechanics, then the laws of nature are statistical laws. Generally, I must repeat my experiment a large number of times, and I shall find that the relative frequencies of outcomes are reproducible, not the individual outcomes themselves. Aside from this fact, it is to be part of my task as a scientist to find out which circumstances are pertinent to the reproducibility of the experiment. As there is no such thing as perfect isolation of my experimental set-up from its surroundings, the reproduction of all variables cannot be exact, but the whole concept of the laws of nature implies that there is no fundamental limit to the degree of accuracy that is attainable.

That there are, however, such limits, any bioscientist will tell us. If he is concerned with epidemiology and if he is called on to assess the efficacy of some public-health measure, he knows that once this measure has been put into effect on a large scale, he cannot hope to reproduce the immunological situation that was present before that application; his sample, humanity as a whole, has been modified irrevocably. As physicists we think of nature as our laboratory, and nature is inexhaustible. No matter how rare is a new elementary particle, we can produce unlimited quantities of precise replicas, given only time and the necessary equipment. There is, however, one major area that is ordinarily considered the legitimate preserve of the physical scientist, in which this principle breaks down. That is the science of the universe as a whole, the discipline that we call cosmology. Though there may be unlimited supplies of rho-mesons, there is an extremely limited supply of universes, — one. If this universe is not in a steady state, — and whether it is no one knows, — then each instant in its history is unique, not to be repeated ever. Hence all processes that are significantly related to the state of the universe as a whole may not be reproducible. The conventional notion of the laws of nature applies to them only in a restricted sense at best.

Let me give you an instance. We know that the population of photons is not in thermodynamic equilibrium, because otherwise the sky should be of uniform brightness, without those extra brilliant spots that we recognize as stars. We have reason to believe that the neutrino flux likewise is far removed from equilibrium, though we are not yet quite able to confirm that conjecture. If there are any particles as yet not known to us (and right now we know four different kinds of neutrinos), they may in turn contribute to the anisotropy of our environment. Suppose now that these particle fluxes affect the elementary particle experiments we perform by some long-range interaction, then the scattering cross sections that we measure might well be dependent on the present state of affairs, and the same experiments performed in a different cosmological period might yield different data. Were we to discover a trend, we should be unable to go back to the previous conditions in order to assure ourselves that the drift that we observe is real and not caused by our improvements in technique. What goes on could ultimately be decided only within the framework of a comprehensive theory that takes into account the relationship between local laboratory experiment and cosmological environment, a theory that in the nature of things could not be tested in all of its aspects.

Astronomers are frequently baffled by this situation. Our most distant observations now enable us to peer into depths of space that correspond, because of the time required for the light to get from there to us, to appreciable fractions of the so-called age of the universe, a little better than half that age, we think. Within a few years it will be possible to run population statistics at that distance and to compare them with corresponding statistics of regions nearby. Suppose we discover significant differences. Does that mean that the universe has evolevd in the meantime, or does it signify geographical differences?

The universe as a whole is intruding on our local observations and experiments yet in another way. If we take an average over a region of the universe large compared to the dimensions of our galaxy but small in terms of cosmological dimensions, that is to say over a region with linear dimensions of the order of 10^7—10^8 light years, then the matter contained in this region possesses a reasonably well-defined state of motion; thus we can define a local frame of reference corresponding to "absolute rest", in contrast to all principles of relativity. True, to a very high order of accuracy the interactions between the constituents of a local mechanical system, such as our solar system, with the surrounding matter at large are negligible compared to the internal interactions of the system. But as our techniques of astronomical measurement are refined, inevitably we shall reach the point where the state of the surrounding matter can no longer be disregarded. Beyond that stage the principles of relativity cannot be tested. In fact, they lose all meaning, for we cannot arrange for experiments to be performed outside our universe. Thus, as we begin to merge cosmological investigations with the normal pursuits of the physicist, we shall encounter ever more numerous instances in which the notion of natural laws will have to be subjected to a critique and, perhaps, be modified.

I shall close with an observation that was made, repeatedly and forcefully, by EINSTEIN. He was concerned with the role of ingenuity in scientific research; he was led to comment on the general philosophy of ERNST MACH, whom he had venerated as a young man and for whom he maintained great intellectual respect throughout his life. Whereas MACH and his school visualized the task of the scientist as that of collecting information by experimentation and by observation, and whereas they viewed scientific theories primarily as convenient means for ordering and structuring the incoming mass of information, EINSTEIN emphasized the role of invention. A collection of sense impressions, taken by itself, adds up to nothing. The human observer must invent concepts that will bring order into the chaos of sense data, and that invention takes place from earliest childhood. These freely invented concepts bear some resemblance to PLATO's ideas. But to become scientific theories, according to EINSTEIN, the new ideas must be tested against observable facts. In other words, not every invention is as valid as every other, and the judgment as to validity must be based on experimentation and observation.

Logical analysis and axiomatization will serve to clarify a situation; conjectures may be based on analogy or on any heuristic principle that will serve. But speculation becomes scientific discovery only to the extent that we succeed in establishing contact with physical reality. Without this contact the range of logical possibilities is too vast to permit us to lay out constructive paths leading to new results. Scientific research is a many-splendored thing, which engages our faculties for experimentation, for dreaming, and for logical analysis. We must cultivate all of our capabilities if we are to reap its fruit.

Chapter 2

The Structure and Content of a Physical Theory

MARIO BUNGE

Departments of Philosophy and Physics, University of Delaware
Newark, Delaware[1]

In analyzing a physical theory we may distinguish at least four aspects of it: the background, the form, the content, and the evidence — if any. By the *background* of a theory we mean the set of its presuppositions. By the *form* or structure, the logico-mathematical formalism quite apart from its reference to physical objects or its empirical support. By the *content* or meaning, that to which the theory is supposed to refer, quite apart from either its form or the way the theory is put to the test. And the *evidence* a theory enjoys is of course the set of its empirical and theoretical supporters. In this chapter an analysis of the internal structure and the external reference of a physical theory will be sketched.

To avoid talking in a vacuum we shall take a particular theory as our object of analysis. And, to make our work more interesting and easier, we shall build the theory from scratch — i.e. with the sole help of generic (logical and mathematical) ideas. We shall proceed in the following way. We shall start with a very simple idea which shall subsequently be expanded into a self-sufficient mathematical formalism F_0. We shall find it impossible to assign a physical interpretation to all the basic concepts of F_0 in order to turn this calculus into a physically interpreted calculus. Whereupon we shall expand this narrow formalism into a more comprehensive structure F_1. Next we shall transform this mathematical theory into a physical theory P_1 by the adjunction of appropriate interpretation assumptions. Then we shall perform a further generalization, yielding a theory P_2 which will turn out to be the well known two component theory of the neutrino. In the process we shall see the self-propelling virtue of mathematics, we shall realize how the mathematical structure limits the possible meanings of the key symbols, and how these meanings are suggested as the theory is worked out rather than being assigned at the outset.

---

[1] Now at the Department of Philosophy, McGill University, Montreal, Canada.

1. The Germ: F_0

Let this be the germ to be developed: A certain process is characterized by the law that the time rate of change of its leading property is balanced by its space rate of change. We do not specify what the process may be nor, a fortiori, what its characteristic property is: like every initial idea ours is vague. But if we can summon the help of mathematics we can make our idea formally precise and we can hope that its logical consequences will provide clues concerning its possible physical meaning. In other words, we shall proceed in a fashion characteristic of contemporary theoretical physics, namely by first building a mathematical structure and then looking around for possible customers fitting that form.

To work. Calling ψ the characteristic property in question, x the position coordinate and t the time coordinate, our hypothesis can be written

$$\frac{\partial\psi}{\partial t}+v\frac{\partial\psi}{\partial x}=0 \tag{1.1}$$

where v is a dimensional constant. To see whether this assumption has any interesting consequences, we integrate it. It is easily checked that any function of the form $\psi(x,t)=f(x-vt)$ solves (1.1). In particular,

$$\psi(x,t)=A\,{\cos \atop \sin}\,k(x-vt) \tag{1.2}$$

and

$$\psi(x,t)=A\,\delta(x-vt) \tag{1.3}$$

where A and k are arbitrary real constants and δ is the Dirac delta.

The first solution can be interpreted as a one-dimensional wave of arbitrary amplitude A and circular frequency $\omega=kv$. The second solution can be interpreted as a point singularity moving with the speed v. This sounds promising, as it suggests a theory that can describe a wave field and an accompanying singularity. In fact, if properly generalized to a three-dimensional space, (1.2) can be interpreted as a plane wave and (1.3) as a point particle, both moving with velocity v. But such as it stands our formalism F_0 has hardly any physical meaning and, moreover, it can be assigned none because physical entities exist in a three dimensional space.

Let us then generalize F_0 in a mathematically trivial but physically significant way, namely by replacing the one-dimensional space underlying (1.1) by a three-dimensional space and correspondingly the scalar v by a vector and $\partial/\partial x$ by the gradient ∇. Furthermore, since the ensuing generalization promises to be interesting, it is advisable to cast it in the form of a lagrangian theory: this will allow us to introduce certain interesting magnitudes. Finally, we shall build the theory in an axiomatic fashion, so as not to miss any important concept and assumption, and so as to facilitate further generalizations. That is, we shall begin by listing the essential (undefined) concepts and by stating the essential (initial) assumptions of the theory, extracting everything else from this foundation. The further concepts will be introduced by definition in terms of the basic (undefined) concepts, and the further statements will be gotten by deduction from the basic (unproved) statements with the help of logic and mathematics. Only a few such consequences will be derived, and this in the usual informal (semi-rigorous) way. The theorems will be helpful in the ulterior work of finding a physical interpretation of the formalism.

2. Axiomatic Foundation of F_1

2.1. Background and Basic Concepts. Like every other theory except logic, ours is based on certain theories. In fact, we take for granted the following two batches of formal (nonfactual) theories: (a) elementary logic (the predicate calculus with identity); (b) analysis, analytic geometry, and all the mathematical theories (e.g., naive set theory and topology) presupposed by the latter. From the stock of concepts this mixed background supplies, we select half a dozen that will play the role of basic concepts: the sets Σ, T, and E^3, the functions ψ and $\mathscr{L}$, and the constant $\boldsymbol{v}$. We call the sextuple

$$B = \langle \Sigma, T, E^3, \psi, \mathscr{L}, \boldsymbol{v} \rangle$$

the concept base of our theory F_1. The latter will boil down to a set of conditions (postulates) B is to satisfy.

We call σ an arbitrary member of Σ; similarly, $t \in T$ and $x \in E^3$. In anticipation of the physical interpretation to be assigned later on, we may read 'σ' as designating (naming) a physical system of the kind Σ; t as a generic instant of time; x as a generic point of the space E^3; ψ as a field coordinate of σ; $\mathscr{L}$ as a lagrangian density; and $\boldsymbol{v}$ as a velocity. But these are extrasystematic remarks: for the time being none of the members of the basic sextuple has a definite physical meaning. The theory will also contain several additional concepts borrowed from or constructed with the help of the theories it presupposes, such as the concept of partial derivative.

2.2. Axioms. The basic statements of the theory are:

A 1 Σ is a nonempty denumerable set.

A 2 T is a nonempty interval of the real line.

A 3 E^3 is the three-dimensional real Euclidean space.

A 4 (a) ψ is a function from $\Sigma \times E^3 \times T$ to the field of complex numbers. (b) For any fixed $\sigma \in \Sigma$, ψ has partial derivatives of the first order for the other four variables.

A 5 (a) $\mathscr{L}$ is a real valued function on $\Sigma \times E^3 \times T$. (b) For every fixed $\sigma \in \Sigma$, $\mathscr{L}$ is twice differentiable w.r.t. all its arguments. (c) $\mathscr{L}$ is integrable over any region of E^3 and any subset of T.

A 6 $\boldsymbol{v}$ is a fixed element of E^3.

A 7 *(Variational principle)*

(a)
$$\mathscr{L}(\sigma, \psi, \psi_t, \psi_x) = \psi \psi_t + \psi (\boldsymbol{v} \cdot \nabla \psi). \tag{2.1}$$

(b) If $\sigma \in \Sigma$ and $t \in [t_1, t_2] \subseteq T$ and $\delta \psi(t_1) = \delta \psi(t_2) = 0$ then, for every σ,

$$\delta \int_{t_1}^{t_2} dt \int_{-\infty}^{\infty} d^3x \, \mathscr{L}(\sigma, \psi, \psi_t, \psi_x) = 0. \tag{2.2}$$

Remark 1. The above axioms characterize roughly the structure or mathematical status of every one of our six basic concepts, with the help of concepts borrowed from the formal background of the theory. The last four axioms bring together some members of the set B of basic concepts, gluing them with logical and mathematical ideas. *Remark 2.* All the above axioms are necessary but they are not equally important. The leading member of the axiom set is *A* 7. All others are subsidiary to it in the sense that they specify the nature of the variables occurring in *A* 7. In an even less formal presentation of the theory, *A* 7 alone would be stated and the remaining assumptions would be at most hinted at. But in an axiomatic treatment every assumption actually used should be stated; in this way it can be kept under control and eventually modified. *Remark 3.* Our axiom set is not phrased in a precisely formalized language but is stated in the naive (informal) way of most of mathematics. The formulation of a completely formalized system would be unbearably lengthy. If only for this reason, applied mathematics and physics will forever utilize the semantic approach, in which one speaks *about* the basic objects of the theory (the members of B) rather than proceeding straight away to stating the relations among all the primitive concepts. *Remark 4.* Since B is a purely mathematical object, the preceding axioms are just mathematical formulas: for the time being they have no physical content although the notation suggests some meaning by association with familiar contexts. In short, the set *A* 1—*A* 7 constitutes a *formalism* or nonfactual theory, namely the foundation of F_1. *Remark 5.* Our axiom system constitutes the basis of a formal theory (in the philosophical not in the metamathematical sense) insofar as it is a mathematical skeleton without any physical meat attached. But F_1 is not an abstract theory, since the postulates specify the nature of every one of the primitives save Σ. In fact Σ, which will eventually be interpreted as the class of physical entities the theory is supposed to describe, is for the time being an abstract set, i.e. a collection of nondescript objects. In other words, F_1 has an interpretation in mathematics and is on this count a far more concrete theory than, say, lattice theory or group theory.

2.3. Some Theorems. Let us now derive the main consequences of the previous assumptions, without giving a thought to mathematical niceties such as the commutativity of the differentiations. In all this section σ shall be kept fixed; in other words, our theorems shall concern an arbitrary member of the future reference class Σ.

Theorem 1. Field equation

$$A\,7 \Leftrightarrow \frac{\partial \psi}{\partial t} + \boldsymbol{v} \cdot \nabla \psi = 0. \tag{2.3}$$

Proof sketch. On performing the first order variations indicated in $A\,7$b we obtain, as the Euler-Lagrange equation corresponding to (2.2),

$$\frac{\partial \mathscr{L}}{\partial \psi} - \sum_{i=1}^{3} \frac{\partial}{\partial x_i} \frac{\partial \mathscr{L}}{\partial \psi_{x_i}} - \frac{d}{dt} \frac{\partial \mathscr{L}}{\partial \psi_t} = 0. \tag{2.4}$$

Now, by $A\,7$b, (2.4) is identical with the second member of the equivalence (2.3).

Theorem 2.

$$\mathscr{L} = 0. \tag{2.5}$$

Proof. By (2.4) and (2.1).

Let us now introduce two abbreviations:

Definition 1

$$\pi =_{df} \frac{\partial \mathscr{L}}{\partial \psi_t}. \tag{2.6}$$

Definition 2

$$\mathscr{H} =_{df} \pi \psi_t - \mathscr{L}. \tag{2.7}$$

In terms of these new (but derived) concepts we state

Theorem 3

(a)

$$\pi = \psi, \tag{2.8}$$

(b)

$$\mathscr{H} = \psi \psi_t. \tag{2.9}$$

Proof. By $A\,7$ and Defs. 1 and 2.

Theorem 4

$$(2.3) \Leftrightarrow \psi_t = \frac{\partial \mathscr{H}}{\partial \pi} = \frac{\partial \mathscr{H}}{\partial \psi}. \tag{2.10}$$

Proof. Differentiate (2.9).

Remark. π is called the momentum density and $\mathscr{H}$ the hamiltonian density of the field ψ. Since $\pi = \psi$, this theory is not strictly hamiltonian. In order to obtain a pair of canonically conjugated field variables we must complicate the lagrangian $A\,7$a. The simplest modification yielding the desired result is replacing ψ by ψ^* in $A\,7$a. This yields $\pi = \psi^*$. Precisely this change must be made in order to quantize the theory in the usual way, but we shall not pursue this line. The point was mentioned because it illustrates the principle "If you want novelty, complicate".

Theorem 5: elementary solutions

(a) If A and the k_i are real numbers, then

$$\psi(x,t) = A \begin{matrix}\cos\\ \sin\end{matrix} \boldsymbol{k}(\boldsymbol{x} - \boldsymbol{v}t). \tag{2.11}$$

(b) If A is a real number, then

$$\psi(x,t) = A\,\delta(\boldsymbol{x} - \boldsymbol{v}t). \tag{2.12}$$

Verification: immediate.

Definition 3

$$H =_{df} \int_{-\infty}^{\infty} d^3x\,\mathscr{H}. \tag{2.13}$$

Theorem 6

$$H = 0. \tag{2.14}$$

Proof. From the odd parity of the integrand.

3. The Physical Theory P_1

We shall now proceed to fill the above structure with a physical content. In other words, we shall interpret the formalism F_1 in physical terms. To this end we must endow the basic concepts of F_1, or at least most of them, with a physical meaning. In other words, the members of the base B must now be regarded as representing either physical things or properties of such. In this way the derived concepts and all of the statements will acquire a physical content.

Let us try the following interpretation of F_1. Assume that Σ is a collection of fields which, for a reason to be given later on, we may call ν-fields. Suppose that T stands for duration and E^3 for space. Finally, assume that ψ represents the field strength, $\mathscr{L}$ the field lagrangian, and $\boldsymbol{v}$ its velocity. These are additional assumptions: they are alien to F_1 as they consist in correspondences between the basic mathematical concepts of the framework F_1 and physical concepts. Since these additional assumptions are not derived but are newly introduced, they are new postulates. Let us state them explicitly:

$A\,8$ Int(Σ) = The collection of ν-fields,
$A\,9$ Int(T) = Time,
$A\,10$ Int(E^3) = Ordinary space,
$A\,11$ Int(ψ) = The ν-field strength,
$A\,12$ Int$(\mathscr{L})$ = The ν-field lagrangian,
$A\,13$ Int$(\boldsymbol{v})$ = The ν-field velocity.

Every one of these statements is an *interpretation postulate* (or hypothesis) rather than an arbitrary designation rule such as, say, "Let E^3 denote the three dimensional Euclidean space". Thus $A\,8$ will turn out to be pointless if, in fact, there is no ν-field in reality; and it will be false if our theory does not correctly describe an entity independently described as a ν-field. On laying down the preceding interpretation assumptions (or semantical postulates) we have completed the *foundations of a physical theory* — which we call P_1 — although we have as yet no idea on how to test it and consequently no idea about its truth value. In other words, we have taken care of the structure and content of P_1 but not of its tests for truth — a methodological problem. This could not be otherwise: before enquiring into the adequacy of a theory this must be at hand. Notice, though, that while the formalism of P_1, i.e. F_1, is described fairly accurately by the axioms $A\,1$ to $A\,7$, the interpretive postulates $A\,8$ to $A\,13$ are quite vague. If they say anything at all it is because the words 'field', 'time', 'space' and the like are meaningful in other chapters of physics. This defect is not peculiar to our theory P_1 but is typical of every physical theory. Indeed, a set of interpretation postulates does no more than *sketch* the meaning of a formalism and moreover it succeeds in performing this modest task only insofar as it establishes a bridge with some other areas of physics: by itself no physical theory has a content.

Let us now read some of the theorems of F_1 in the light of the interpretation postulates $A\,8$ to $A\,13$: this will sharpen the semantic profile of P_1. Equation (2.3) is the ν-field equation, i.e. the formula that specifies ("defines" in the usual jargon) the structure of the thing ν our theory P_1 is supposed to refer to and even describe. Theorem 5 may be read as follows: the simplest system ν consists of two entities: (a) a plane wave of amplitude A and frequency $\omega = kv$ travelling with velocity $\boldsymbol{v}$, and (b) a singularity moving at the same speed, i.e. accompanying the wave. The remaining theorems have so far no obvious physical meaning. For instance, we cannot interpret H as the field energy if only because there is no equation for π_t to match (2.10), other than (2.10) itself. Indeed, as will be recalled, $\pi = \psi$. Also, no density-current vector can be defined unless complex field components are added, whence our ν-field is electrically neutral.

In summary, P_1 is a unitary theory describing at the same time a field and a singularity in it, which may be regarded as a particle associated with the field. Actually the interpretation of a field singularity as a particle is justified only insofar as both are localized. The other essential property of particles, namely mass, is absent from our theory, which is purely kinematical.

Let us next generalize P_1. Upon a purely mathematical expansion, P_1 will be transformed into a richer theory P_2 which happens to account for a real entity — the neutrino. What began as a game will end up in a serious piece of science.

4. P_2: The Neutrino Theory

P_1 can be generalized in a number of directions. One possible move is to increase the number of field components by postulating ψ to be proportional to a $n\times 1$ matrix of complex functions; this will require $\boldsymbol{v}$ to be a $n\times n$ matrix. That is, we may postulate the lagrangian density

$$\mathcal{L}=\psi^+\psi_t+\psi^+c(\boldsymbol{\sigma}\cdot\nabla\psi) \tag{4.1}$$

with $\psi^+\equiv\tilde{\psi}^*$ and $c\in R$. The corresponding system of field equations is then

$$\frac{\partial\psi}{\partial t}+c(\boldsymbol{\sigma}\cdot\nabla\psi)=0. \tag{4.2}$$

So far, the order n and the representation of the σ_i are not specified. The simplest choice is $n=2$, in which case they cannot be other than the Pauli 2×2 matrices:

$$\sigma_1=\begin{pmatrix}0&1\\1&0\end{pmatrix},\qquad \sigma_2=\begin{pmatrix}0&-i\\i&0\end{pmatrix},\qquad \sigma_3=\begin{pmatrix}1&0\\0&-1\end{pmatrix}. \tag{4.3}$$

In this case the (4.2) abbreviate a system of two equations known as WEYL'S equations — the nucleus of the two component theory of the neutrino. (This theory was born in 1929, buried in 1933 and resurrected in 1956.) The deduction of a few consequences will shed some light on the possible physical meaning of the new formalism F_2 developed so far, or rather on the meaning that can possibly be attached to it.

To begin with, by a suitable transformation all but the third term of the scalar product $\boldsymbol{\sigma}\cdot\nabla\psi$ can be eliminated, leaving

$$\frac{\partial\psi}{\partial t}+c\,\sigma_3\frac{\partial\psi}{\partial x_3}=0. \tag{4.4}$$

This system has two elementary solutions:

$$\psi_{1,2}=A_{1,2}\,{\cos \atop \sin}\,k(x_3\mp ct) \tag{4.5}$$

and

$$\psi_{1,2}=A_{1,2}\,\delta(x_3\mp ct) \tag{4.6}$$

with A_1 and A_2 complex numbers. Possible interpretation: the theory refers to (models, mirrors, represents) two sets of waves and their accompanying field singularities, which move in opposite directions along the forward cone $x_3=\pm ct$.

Next we fourier-analyze the field equations (4.2), obtaining

$$[-\omega+c(\boldsymbol{\sigma}\cdot\boldsymbol{k})]\,\varphi(\boldsymbol{k})=0,\quad \text{with}\quad \varphi=\mathcal{F}[\psi]. \tag{4.7}$$

For nontrivial solutions the circular frequency is

$$\omega=ck\,\hat{\sigma} \tag{4.8}$$

where

$$\hat{\sigma} =_{df} \frac{(\boldsymbol{\sigma} \cdot \boldsymbol{k})}{k} \tag{4.9}$$

is called the helicity. For a given value k of the wave momentum there are two $\hat{\sigma}$ states: (a) $\hat{\sigma} = +1$ for $\boldsymbol{\sigma}$ parallel to $\boldsymbol{k}$, and (b) $\hat{\sigma} = -1$ for $\boldsymbol{\sigma}$ antiparallel to $\boldsymbol{k}$. A further development of the theory shows that the value $\hat{\sigma} = +1$ can be made to correspond to a right-handed neutrino and $\hat{\sigma} = -1$ to a left-handed neutrino.

Several other generalizations are thinkable. For instance, we can quantize P_2 since we can now identify pairs of canonically conjugate variables: ψ_α and ψ_α^*. The quantization will consist in representing the field components by operators and in adding the postulate that the commutators (or the anticommutators) of the field components equal a certain quantity. This will not be attempted here for we have achieved what we wanted: a couple of theories that, though mathematically quite simple, have an interpretive haziness about them that is comparable to any of the best scientific theories — notwithstanding the textbook image according to which every symbol occurring in a scientific theory has a perfectly definite meaning assigned by an operational "definition", whatever this may mean.

5. Analysis

Let us retrace our steps. We began by toying with a simple idea referring to a possible physical fact. This idea was extremely rudimentary i.e. imprecise both in form and in content. In order to make it more definite we did not attempt to make any measurements, because the very design and interpretation of a measurement requires clear ideas. Instead we restated our idea in a mathematical way [equation (1.1)]. We then performed a preliminary exploration of the logical consequences of our initial idea and this gave us a clue concerning the physical content it might be filled with: in fact we found two theorems [formulas (1.2) and (1.3)] that, properly extended, might be assigned an interesting interpretation — namely, as referring to a wave-particle pair. But this required a prior mathematical extension of our idea.

In order to enjoy the freedom of mathematics — so as to expand our initial idea — we forgot temporarily about physical meanings and proceeded to generalize the initial idea in a purely mathematical way: we replaced the product of two scalars by the scalar product of two vectors, thus ending up with equation (2.3). What is more, in anticipation of future developments we placed the new egg in the ready-made nest of Lagrangian "dynamics" hoping that this big bird, which cares not for the feather, would do the hatching. That is, the initial idea, once generalized, was restated in the form of an action principle. Once this was obtained, the whole process was reversed: we gave an axiomatic reconstruction of the theory which, in actual fact, had grown spontaneously. To this end we examined the leading hypothesis — the action principle — in search for the basic concepts. These turned out to be the members of the sextuple B — the concern of the theory. We then went on to state the subsidiary assumptions A 1 to A 6 that, by characterizing those building stones, would provide the proper background for the leading axiom A 7. In this way the axiomatic foundations of a formal theory F_1 were built. The rest was all built upon it: the new concepts were defined in terms of the basic concepts of F_1 and of other, nonspecific concepts, and the new statements were derived from the explicitly stated assumptions alone with the help of logic and mathematics. But the choice of the concepts to be defined and of the theorems to be derived was of course guided by physical intuition, i.e. by our experience concerning the possible physical meaning and significance of the ideas in play.

In short, we first had a theory *construction* process, then the reverse process of theory *reconstruction* or logical organization. In both cases we availed ourselves of existing mathematical ideas; in the case of the axiomatic reconstruction we also used a general matrix for the gestation of physical theories — Lagrangian "dynamics" — and some metamathematical ideas concerning axiom systems. In no case did we use anything like induction, said to be the motive force of science: the original egg was invented, its generalization was suggested — almost forced — by mathematics, its restatement as a variational principle was an elementary exercise in guesswork, the discovery of the axioms subsidiary to it was plain mathematical common sense, and the derivation of theorems was sheer informal deduction.

But having built F_1 we were still midway: mathematical formalisms, the goal of mathematical work, are just means for the physicist, who wants physically interpreted calculi. The structure F_1 had to be given an interpretation and, more precisely, a physical one. This was done by correlating the basic symbols of B with physical objects such as fields and their characteristic traits. That is, we added the six initial hypotheses A 8 to A 13, which made explicit the interpretation we had in mind. In this way a physical theory proper, P_1, was born. This interpretation of the formalism F_1 was suggested by analogy with other theories: analogy, rather than either induction or deduction, was here the driving force.

This possible physical model P_1 of the formalism F_1 turned out to be neither unique nor a fully determinate interpretation. It is not *unique*: in fact, alternative interpretations of F_1 are conceivable. For example, ψ could also be interpreted as the density of a fluid flowing with speed $\boldsymbol{v}$.

And P_1 is not a *fully interpreted* system for the following reasons. Firstly, the lagrangian density does not represent a definite physical property but is rather a collective representative of all the properties of the system concerned; if preferred, it is a source property from which every single property can be derived. Second, the chief concept of them all, i.e. that of ν-field — symbolized by Σ — remains indeterminate except for P_1 itself: unlike other physical concepts, such as those of time and mass, which occur in many other theories, this one is a very specific concept and moreover one determined by a rather poor theory. This trait — the *partial interpretation* of the formalism of our theory — is common to all high-brow scientific theories and it is bivalent: on the one hand it encourages a wild proliferation of *ad hoc* (unwarranted) interpretations, while on the other it leaves us freedom to pick up the right referent. Remember that the Yukawa theory was first assigned the μ-meson, later on the π-meson. In short, the reference set of a physical theory need not be perfectly definite for the theory to be valuable. The theory itself may be instrumental in identifying its own referent.

Look again at the interpretation postulates $A\,8$ to $A\,13$. Although these assumptions do stipulate a physical meaning — if only sketchily — they indicate no set of empirical operations. For instance, $A\,9$ does not state that T is a set of clock readings (mathematically an insignificant subset of fractionary numbers) but a set of instants of time, whether any of them are measured or not. Neither of our theories, P_1 and P_2, establishes the way they can be put to the test. To suggest this would be the task of further theories, linking the unobservable ν-field with observable macrofacts. That the test of a theory should require the cooperation of additional theories is not peculiar to P_1 and P_2 but is a general feature. Thus not even rigid body mechanics, which after all refers to observable pieces of matter, supplies all that is necessary for its own test: the latter requires, in addition, a fragment of optics and another of gravitation theory since both light and gravity are essential ingredients in the experimental tests of the theorems of rigid body mechanics. In any case, the meaning of a physical theory is not established by relating it to test situations, which are never realistically accounted for by a single theory. Observations, measurements and experiments are conducted to find out the truth value or degree of factual adequacy of a theory, not to endow in with a content. Tests are neither necessary nor sufficient to secure meaning; on the other hand the possible meaning of a theory must be proposed, if only in outline, before the test of it can be planned and executed.

What about the empirical test of our theories P_1 and P_2? Clearly, P_1 is too accomodating: it is confirmed, e.g., by any scalar wave propagation process. When accompanied by such a lack of specificity, large coverage is not too valuable. On the other hand P_2 fares better as regards testability because, in addition to pointing to a sort of wavicle traveling spontaneously with constant velocity c, it assigns this entity the property of helicity [equation (4.9)], which can be positive (right-handed screw) or negative (left-handed screw). And such an entity has actually been "observed": it is the neutrino, sought and found thanks to the theory. Yet the testability of P_2 is not impressive, as was to be expected since it is a purely kinematical theory that assigns the ν-field neither a mass nor a charge and that does not commit itself to any mechanism for the interaction between neutrinos and other entities. In order to enhance the testability of P_2 this theory must be enlarged to cover such interactions. Which is a nice illustration of yet another nonempiricist thesis, namely that the testability of a theory is not only a question of experimental technique and that it can be enhanced by generalization rather than hampered by it.

6. Conclusion

To conclude: a physical theory is a mathematical formalism endowed with a physical meaning and therefore susceptible to empirical tests, eventually with the assistance of further theories. The formalism is a hypothetico-deductive system, i.e. a set of formulas generated by applying logic and mathematics to a set of initial assumptions. These basic hypotheses or axioms are given some physical meaning by interpreting the basic symbols of the theory in physical terms, i.e. by adding interpretation postulates that make those symbols stand for definite physical things and properties thereof. One and the same formalism can be attached several interpretations, a process whereby several physical theories can emerge.

Foundational work is as much concerned with digging out the background of theories as with exhibiting their structure and their meaning. But in practice the interpretation problem is the more important precisely because it cannot be solved in an exact way. Also, it is rarely handled in a thorough and consistent way — or if it is so handled then it is done in the light of the operationalist myth according to which every symbol has to stand for a set of laboratory operations. On the other hand one can always hope that the mathematician will tidy up the sloppy form of an otherwise valuable physical theory. Once the importance of interpretation problems is realized and it is understood that proposing a reinterpretation of a given formalism is tantamount to producing a new physical theory — even if no new mathematical formulas are offered and no new experiments are immediately suggested — the creative aspect of foundational research can be appreciated alongside its indispensable critical function.

Our analysis has been extremely incomplete. Thus we have not mentioned that every hypothetico-deductive system is a filter; we have made no use of the calculus of deductive systems; we have surmised that the basic concepts and initial assumptions of our theories are independent without subjecting this conjecture to any tests; we have not even mentioned that T and E^3 remain largely indeterminate unless they are elucidated by a theory of time and a physical geometry respectively; we took for granted that Lagrangian "dynamics" is a framework for physical theories rather than a specific physical theory; we have failed to mention that the problem of interpreting an abstract theory within mathematics has given rise to a whole new branch of metamathematics — model theory — whereas the semantics of physical theories is hardly in the making.

In conclusion the subject of this chapter in an intriguing field, particularly underdeveloped on the semantical side, that can now be advanced in the light of the logical, algebraic, and metamathematical studies undertaken by mathematicians — which studies are however insufficient for our purposes, as a physical theory is not interpreted in another theory but by reference to the external world. Mathematicians have learned long ago the value of analyzing, cleansing, and orderly reconstructing what has been built intuitively. When shall we learn it?[1]

REFERENCES

Bunge, M.: Scientific research. Vol. I: The search for system. Vol. II: The search for truth. Berlin-Heidelberg-New York: Springer 1967.

— Foundations of Physics. Berlin-Heidelberg-New York: Springer 1967.

Carnap, R.: Introduction to symbolic logic and its applications. New York: Dover 1958.

Henkin, L., P. Suppes, and A. Tarski (Eds.): The axiomatic method. Amsterdam: North-Holland 1959.

Stoll, R. R.: Set theory and logic. San Francisco and London: W. H. Freeman and Co. 1963.

Suppes, P.: Introduction to logic. Princeton, N. J.: Van Nostrand 1957.

Tarski, A.: Logic, semantics, metamathematics. Oxford: Clarendon Press 1956.

[1] I am thankful to Dr. Walter Felscher (Mathematisches Institut der Universität Freiburg) for a critical reading of the MS.

Chapter 3

Space-Time Structures in Classical Mechanics

WALTER NOLL

Department of Mathematics, Carnegie Institute of Technology
Pittsburgh, Pennsylvania

1. Introduction

The English language contains many words that denote spatial or temporal concepts: 'now', 'later', 'soon', 'simultaneous', 'here', 'there', 'far', 'location', 'equidistant', etc. The grammar is in part organized in accordance with temporal categories: present, past, future. If we tried to remove all words, prefixes, and suffixes with a temporal or spatial meaning from the language we would surely all but destroy it. The system of temporal and spatial concepts of a natural language such as English constitutes a *verbal space-time structure*. It is not a very precise system, but it serves very well as a framework for the common experiences of human life.

The geometry of the ancient Greeks and the spatial and temporal concepts of the mechanics of GALILEO and NEWTON may be viewed as being refinements of the intuitive verbal space-time structure, refinements which resulted in a very precise mathematical system. This system, which I call the *classical space-time structure*, provides the basis for several very successful branches of physics, chief among them the mechanics of particle systems of NEWTON and the mechanics of rigid bodies of EULER. In Sect. 2 I shall give a brief outline of a modern version of the classical space-time structure.

Until the beginning of the 19-th century there were very few people, if there were any, who could even imagine a system that might replace classical space-time. KANT, for example, regarded the valid statements of classical geometry and mechanics as being "a priori" and "synthetic"; i.e., he considered them to be truths about reality not derived from experience and yet not mere tautologies. This judgement reflects the view that the classical space-time structure is not just an expedient framework for physical experience, but is indeed the only conceivable such framework. It is a very understandable view, because before the invention of non-Euclidean geometry it must have been impossible to imagine how another space-time structure might be formulated.

A non-classical space-time structure cannot easily be described entirely in words, because words have connotations that imply the classical structure. The term "space-time" itself has misleading connotations when applied to a nonclassical system. The natural language keeps us in the prison of classical space-time. The only language within which a non-classical structure can be unambiguously formulated is the language of axiomatic mathematics. The methods of modern mathematics make it possible to fabricate almost at will structures that can play the role of classical space-time.

One may ask why anybody would wish to consider a nonclassical space-time at all. The reason is, of course, that the classical structure, while adequate as a basis for the concepts of ordinary experience and of the older branches of physics, is inadequate for some of the newer physical disciplines, inadequate, in particular, for relativistic and quantum physics. These disciplines require space-time structures that radically deviate from the classical one. It is not the purpose of this lecture to elaborate on these structures. Rather, I here content that classical space-time is not well suited even for some venerable branches of mechanics, and I shall develop another structure which I believe to be more appropriate for these branches. This structure, which I call *neo-classical space-time*, will be presented in Sect. 3; and in Sects. 4—6 I shall indicate how it can be used for developing classical mechanics.

Many of the concepts of classical space-time reflect the fact that we humans live on this solid earth, which in daily life is always available as a frame of reference for specifying "locations". The first blow to the classical system was dealt by COPERNICUS, who deprived the earth of its once secure place as the universal frame of reference. The concept of a "location" in interstellar space is much more problematical than that of a location on earth. NEWTON's *absolute space*, as the set of all possible locations, has since its inception been regarded with unease by most thinkers. Nonetheless, the "Newtonian" mechanics built upon this concept of absolute space has been very successful. Actually, use of the concept of absolute space is one among several ways of accounting for the phenomenon of inertia. Classical space-time with its absolute space has been most successful in those branches of mechanics in which inertia plays the central role. Such is not the case, however, in many of the branches of the mechanics of continuous media, where inertia is often of minor importance or sometimes even altogether neglected. In these branches of mechanics absolute space is an artificial and inappropriate concept. If it is used anyway, it is necessary to compensate for its arbitrariness by introducing a requirement of invariance, called the *principle of frame-indifference or objectivity* (c.f. [1], Sect. 19). The description of mechanics in terms of the neo-classical space-time, which has no room for an absolute space, shows clearly why this principle is needed.

2. The Event-World of Classical Space-Time

The *absolute space* is a set E consisting of *points (locations)* $x, y, \ldots$. The set E is endowed with a mathematical structure defined by a *distance function* δ which associates with each pair x, y of points a number $\delta(x, y)$, *the distance from* x *to* y. The distance function δ is subject to certain axioms which ensure that δ is a Euclidean metric that gives E the structure of a *Euclidean space*. This Euclidean structure makes it possible to define a unique *translation space* V_E of E, which is a vector space with inner product consisting of automorphisms of E. (For details, see [2], Sect. 4.)

The *event-world* of classical space-time is the set $W = E \times R$ of all pairs (x, t), where $x \in E$ and $t \in R$, the set of real numbers. The point x is called the *location*, the number t the *time* of the event (x, t).

3. The Event-World of Neo-Classical Space-Time

The *event-world* of neo-classical space-time is a set W consisting of *events* $e, f, \ldots$. The set W is endowed with a mathematical structure defined by a *time-lapse function* τ and a *distance function* δ, subject to the axioms $(T_1)-(T_4)$ and $(D_1)-(D_3)$ stated below.

(T_1) The time-lapse function τ assigns to each pair e, f of events a number $\tau(e, f)$, called the *time-lapse* between e and f.

(T_2) For any $e, f \in W$,

$$\tau(e, f) = -\tau(f, e). \tag{3.1}$$

(T_3) For any $e, f, g \in W$,

$$\tau(e, f) + \tau(f, g) = \tau(e, g). \tag{3.2}$$

(T_4) For any $e \in W$ and any $t \in R$ there is a $f \in W$ such that $\tau(e, f) = t$.

We say that e is *earlier* than, *later* than, or *simultaneous* with f according to whether $\tau(e, f) > 0$, < 0, or $= 0$.

The set of all pairs of simultaneous events

$$S = \{(e, f) \mid \tau(e, f) = 0\} \tag{3.4}$$

is an equivalence relation on W, as can easily be seen to follow from (T_2) and (T_3). This equivalence relation determines a partition Γ of W into classes T of simultaneous events such that

$$S = \bigcup_{T \in \Gamma} T \times T. \tag{3.5}$$

The equivalence classes T will be called *instantaneous spaces* or simply *instants*. If $e \in T$ we say that *the event e happens at the instant T*.

The value of the time-lapse $\tau(e, f)$ depends only on the instants T and S at which e and f happen. Therefore, we can define unambiguously a time-lapse

$$\bar{\tau}(T, S) = \tau(e, f) \quad \text{if} \quad e \in T, f \in S \tag{3.6}$$

between two instants T and S.

(D_1) The distance function δ assigns to each pair $(e, f) \in S$, i.e., to each pair of simultaneous events, a number $\delta(e, f)$, called the *instantaneous distance* between e and f.

(D_2) For each instant T, the restriction δ_T of δ to $T \times T$ is a Euclidean metric on T.

To say that δ_T is a Euclidean metric means that it gives to the instant T the structure of a Euclidean space. The translation space of T will be denoted by V_T.

(D_3) For each instant T, the dimension of the translation space V_T is 3.

Physically, the values $\tau(e, f)$ of the time-lapse function τ are to be interpreted as the results of time-measurements with clocks. The axioms (T_2)—(T_4) reflect familiar experiences with such measurements. The values $\delta(e, f)$ of the distance function δ are to be interpreted as the results of distance measurements with measuring sticks. The value $\delta(e, f)$ is defined only when e and f are simultaneous because each distance measurement is made at a particular instant. The axioms (D_2) and (D_3) are the abstract of thousands of years of experience with distance measurements.

An *automorphism* α of the event-world W is a one-to-one mapping of W onto itself which preserves time-lapses and distances. Thus, an automorphism α satisfies

$$\tau(\alpha(e), \alpha(f)) = \tau(e, f) \quad \text{for all} \quad e, f \in W \tag{3.7}$$

and

$$\delta(\alpha(e), \alpha(f)) = \delta(e, f) \quad \text{for all} \quad (e, f) \in S. \tag{3.8}$$

If $U \subset W$ we write

$$U^\alpha = \{\alpha(e) \mid e \in U\} \tag{3.9}$$

for the set of all images under α of events in U. In this way, the mapping $T \to T^\alpha$ of Γ onto itself is an automorphism of Γ in the sense that it preserves the time-lapse function $\bar{\tau}$ defined by (3.6).

An automorphism α of W also induces isomorphisms $V_T \to V_{T^\alpha}$ between the translation spaces of the instantaneous spaces.

4. Material Universes, Motions

A *material universe*[1] is a set $\mathcal{U}$ consisting of *bodies* $B, C, \ldots$. It is assumed that $\mathcal{U}$ is partially ordered by a relation $<$, and we say that C *is a part of* B if $C<B$. The body B is said to be *separate* from the body C if B and C have no part in common, i.e., if there is no $D\in\mathcal{U}$ such that both $D<B$ and $D<C$.

The material universe appropriate to a system of discrete particles consists of the collection of all subsets of a finite set, whose members represent the particles. In this case, the relation $<$ is the set-inclusion. For continuum mechanics, however, more complicated material universes must be considered.

A *motion* of the material universe $\mathcal{U}$ is a function M which assigns to each body $B\in\mathcal{U}$ a subset $M(B)$ of W such that

$$M(B)\subset M(C) \quad \text{if} \quad B<C \tag{4.1}$$

and

$$M(B)\cap T\neq\emptyset \quad \text{for each} \quad T\in\Gamma. \tag{4.2}$$

The set $M(B)$ is called the *set of events experienced by* B during the motion M, or simply the *world-tube* of B.

The requirement (4.1) states that a part experiences fewer events than the whole, and (4.2) expresses the fact that bodies cannot appear out of nothing nor disappear into nothing.

An automorphism α of the event-world W induces a transformation $M\to M^{\alpha}$ on motions; it is defined by

$$M^{\alpha}(B)=\big(M(B)\big)^{\alpha} \quad \text{for all} \quad B\in\mathcal{U}. \tag{4.3}$$

The above is only the beginning of a kinematics based on the neoclassical space-time. A more detailed development will be presented in future publications.

5. Force Systems, Dynamical Processes

A *force system* for a material universe $\mathcal{U}$ is a function φ which assigns to every triple (B, C, T), where B and C are separate bodies in $\mathcal{U}$ and $T\in\Gamma$ is an instant, a vector $\varphi(B, C, T)$ in the translation space V_T of T. The value $\varphi(B, C, T)$ is called the *force exerted by the body* C *on the body* B *at the instant* T. Force systems are subject to restrictions which will not be stated here. (They are similar to the ones given in [*3*], Sect. 3, and [*4*], Sect. 4.)

---

[1] The concept of material universe used here differs from the one of references [*3*] and [*4*]. In these papers, bodies are assumed to be certain subsets of a universal set, while here they need not be sets at all.

A *dynamical process* for the material universe $\mathcal{U}$ is a pair $\Pi = (M, \varphi)$, where M is a motion of $\mathcal{U}$ and φ a force system for $\mathcal{U}$, such that the fundamental laws of balance of forces and moments are statisfied. A precise statement of these laws in the present framework and a detailed treatment of dynamics will be given in future publications. The totality of all dynamical processes for $\mathcal{U}$ will be denoted by $\mathcal{D}$.

An automorphism α of W induces a transformation $\varphi \to \varphi^\alpha$ on force systems; it is defined by

$$\varphi^\alpha(B, C, T^\alpha) = \alpha(\varphi(B, C, T)), \tag{5.1}$$

where the right hand side is the image under the isomorphism $V_T \to V_{T\alpha}$ induced by α. The automorphism α induces the transformation

$$\Pi = (M, \varphi) \to \Pi^\alpha = (M^\alpha, \varphi^\alpha) \tag{5.2}$$

on the set $\mathcal{D}$ of all dynamical processes, where M^α is defined by (4.3) and φ^α by (5.1). The fundamental laws of balance are invariant under transformations of the form (5.2), so that dynamical processes are transformed into dynamical processes.

In order to have consistency of the theory described here with the conventional approaches to classical mechanics one must include inertial forces in the force systems φ on an equal footing with other kinds of forces (cf. [*3*]).

6. Constitutive Classes

The nature of many problems in mechanics can roughly be described as follows: among all conceivable dynamical processes for a material universe, select the one that will actually occur. In order to make such a selection, one must know something about the particular material properties of the bodies which belong to the material universe. Conventionally such properties are described by "force laws", "stress-strain relations", or similar types of constitutive laws.

A way of making precise the concept of a material property within the present framework is that of using the notion of a constitutive class: A *constitutive class* for a pair B, C of separate bodies is a subset $\mathcal{C}(B, C)$ of the set $\mathcal{D}$ of all dynamical processes, subject to the following requirements:

(I) If $\Pi_1 = (M_1, \varphi_1)$ and $\Pi_2 = (M_2, \varphi_2)$ are two dynamical processes such that

$$M_1(D) = M_2(D) \tag{6.1}$$

for all parts $D < B$ or $D < C$ and

$$\varphi_1(B, C, T) = \varphi_2(B, C, T) \tag{6.2}$$

for all instants $T \in \Gamma$, then $\Pi_1 \in \mathcal{C}(B, C)$ if and only if $\Pi_2 \in \mathcal{C}(C, B)$.

(II) $\mathscr{C}(B, C)$ is stable under automorphisms; i.e., if $\Pi \in \mathscr{C}(B, C)$ and if α is an automorphism of W, then $\Pi^\alpha \in \mathscr{C}(B, C)$, where Π^α is defined by (5.2).

The first of these requirements is a principle of irrelevance, stating that the material properties of the bodies B and C concern only B and C, and not anything else in the universe. The second requirement is a principle of homogeneity for the event-world, expressing the condition that events may have no individuality beyond the one conferred to them by motions of material universes.

In conventional treatments of mechanics material properties are defined by means of *constitutive equations*. Such constitutive equations actually define classes of dynamical processes in the sense of Sect. 5 if a suitable concept of *frame of reference* is employed. One can show that these constitutive equations must satisfy the principle of material frame-indifference and the principle of local action (see [1], Sect. 23) if they are to define constitutive classes in the sense described above.

REFERENCES

[1] TRUESDELL, C., and W. NOLL: The non-linear field theories of mechanics. In: Handbuch der Physik (S. FLÜGGE, ed.), vol. III/3. Berlin-Göttingen-Heidelberg: Springer 1965.

[2] NOLL, W.: Euclidean geometry and Minkowskian chronometry. Am. Math. Monthly **71**, 129—144 (1964).

[3] — La Mécanique Classique Basée Sur Un Axiome D'Objectivité, Colloque internat. sur la méthode axiomatique dans les mécaniques classiques et nouvelles 1959, p. 47—56. Paris 1963.

[4] — The foundations of classical mechanics in the light of recent advances in continuum mechanics. Proc. of the Berkeley Symposium on the Axiomatic Method, p. 266—281, Amsterdam 1959.

Chapter 4

Foundations of Continuum Mechanics[1]

Clifford A. Truesdell

Department of Mechanics, The Johns Hopkins University
Baltimore, Maryland

Last week Mr. Noll presented a new, compact, and general theory of space-time structure for Euclidean continuum mechanics. In mathematical science organization can be effected at various levels. For example, we may construct the real numbers from the integers, or we may take axioms for the real numbers themselves as our starting point. Mr. Noll's space-time structure furnishes a foundation for the theory of constitutive equations he formulated some years ago, but it does not change that theory or render it either more or less precise. Along with deepening the foundations, mathematical science strives also to broaden the structure. The great clarity gained from the abstract approach to constitutive equations for purely mechanical phenomena has made it possible to extend the structure of rational mechanics so as to include thermo-energetic effects in comparable generality. Thermostatics, now a century old, was never intended to apply to problems of deformation and motion; the linear "irreversible thermodynamics", which rests on applying classical thermostatics to volume elements, has fallen so far behind recent views on mechanics as to put it quite out of the running, and the new thermodynamics, developed by Mr. Coleman, makes no use of its concepts or apparatus.

---

[1] Acknowledgement. Some of the later paragraphs of this lecture follow nearly verbatim some earlier publications:

"Rational Mechanics of Deformation and Flow" (Bingham Medal Address), Proceedings of the Fourth International Congress of Rheology (1963), Interscience Publishers, New York, 1965, Volume 2, pp. 3—30.

"The Modern Spirit in Applied Mathematics", ICSU-Review of World Science, Elsevier Publishing Company, Amsterdam, 1964, Volume 6, pp. 195—205.

"Method and Taste in Natural Philosophy", Six Lectures on Natural Philosophy, Berlin-Heidelberg-New York: Springer 1966, pp. 83—107.

For permission to reprint these passages I thank the editors and publishers of the above publications.

In this lecture I wish to outline the theory of thermo-mechanics as it appears today, with no claim of generality sufficient to include all current proposals, let alone of finality. Rather, COLEMAN'S thermodynamics corresponds to NOLL'S theory of purely mechanical phenomena in simple materials, operating at the same level of precision and abstractness.

The basic principles of mechanics and energetics are taken as the integral principles of balance which lead to the following field equations and field inequality:

Conservation of mass:

$\dot{\varrho}+\varrho \operatorname{div} \dot{\boldsymbol{x}}=0,$ $\qquad \varrho=$ mass density,

$\dot{\boldsymbol{x}}=$ velocity.

Balance of linear momentum:

$\operatorname{div} \boldsymbol{T}+\varrho \boldsymbol{b}=\varrho \ddot{\boldsymbol{x}},$ $\qquad \boldsymbol{T}=$ stress tensor,

$\ddot{\boldsymbol{x}}=$ acceleration,

$\boldsymbol{b}=$ body force density.

Balance of moment of momentum:

$\boldsymbol{T}=\boldsymbol{T}^T.$

Balance of energy:

$\varrho \dot{\varepsilon}=\operatorname{tr}(\boldsymbol{T}\boldsymbol{D})+\operatorname{div} \boldsymbol{h}+\varrho q,$ $\qquad \varepsilon=$ internal energy,

$\boldsymbol{D} \equiv \frac{1}{2}\left(\nabla \dot{\boldsymbol{x}}+(\nabla \dot{\boldsymbol{x}})^T\right)=$ stretching tensor,

$\boldsymbol{h}=$ heat flux vector,

$q=$ heat absorption density.

Growth of entropy:

$\varrho \dot{\eta} \geqq \operatorname{div}\left(\frac{\boldsymbol{h}}{\theta}\right)+\varrho \frac{q}{\theta},$

$\theta=$ temperature, assumed always positive,

$\eta=$ specific entropy.

Of the fields occurring here, one set consists in quantities defined in terms of the motion and the mass of the body: ϱ, $\dot{\boldsymbol{x}}$, $\ddot{\boldsymbol{x}}$, $\boldsymbol{D}$. The remainder, consisting in $\boldsymbol{T}$, $\boldsymbol{b}$, ε, $\boldsymbol{h}$, q, η, and θ are primitive variables. Various physical interpretations for them all are well known, and I see no reason to give any further explanation for the two last in particular.

The last two conditions could well be called "the first and second laws of thermodynamics", were it not for the truly magnificent clouds of unreason to which even a whisper of these terms gives rise. Every physicist knows exactly what the first and second laws mean, but it is my experience that no two physicists agree about them. We are interested here in specific theory, not talk, so we may leave the physics to the physicists and stay with our equations.

In a typical problem of continuum mechanics, the body force $\boldsymbol{b}$ and the heat absorption q are specified as part of the data. Generally both are zero, but the fields of gravity and thermal radiation also occur frequently, and in addition there may be force and heating due to electric currents. I mention these examples because they show that $\boldsymbol{b}$ and q may be quite various. Indeed, when we start to think about what restrictions ought to be imposed upon $\boldsymbol{b}$ and q, we cannot find any in principle. If $\boldsymbol{b}$ and q are assigned, the equations of linear momentum and energy restrict the values of the other fields occurring in them. If, however, we ask for the totality of all conceivable processes, we see that if any smooth motion and stress field whatever are given, the equation of linear momentum determines a body force $\boldsymbol{b}$ such as to balance their effects; if, in addition, we prescribe any smooth internal energy and heat flux fields, the equation of energy determines a heat absorption q such as to balance them, too. That is, provided the stress tensor be symmetric, the principles of mass, momentum, and energy impose *no restrictions whatever* on the class of histories of deformation and temperature.

The principle of growth of entropy, called the "Clausius-Duhem inequality", is of a different kind. Since it is an inequality, it determines no new quantity uniquely. If we suppose the temperature field be given, the principle of entropy growth *lays down a prohibition* on the growth of the entropy field. The entropy, unlike the other fields of thermo-mechanics, *cannot be assigned arbitrarily* even by the most artificial adjustment of forces and supplies of heat. This observation is the key to the new thermodynamics.

So much for the general principles. Highly indeterminate, they provide us with no more than a framework to be clothed with particular, illustrative concepts of material response. The variety of materials is expressed by *constitutive equations*. The common experience that forces are required in order to deform a body and that heat must be supplied in order to effect a change of temperature, or, in older terms, that deformation "causes" internal forces and temperature differences "cause" flow of heat, is expressed as the following two statements:

$\boldsymbol{T}$ is determined by $\boldsymbol{\chi}$, where $\boldsymbol{x}=\boldsymbol{\chi}(X, t)$ is the *motion*, namely, the mapping of the particle X into the place $\boldsymbol{x}$ at the time t,

$\boldsymbol{h}$ is determined by grad θ.

The "causes" are χ and grad θ, and the "effects" are $\boldsymbol{T}$ and $\boldsymbol{h}$, respectively. Then there is the internal energy ε and the entropy η (which, if I dared, I should call "heat"); they, too, are effects of deformation and temperature change. We are thus faced by statements of *determinism,* relating various causes to various effects. Which causes give rise to which effects? Physical experience, already formalized in a dozen or more special theories of various degrees of seniority, does not give a definite answer. When in doubt, be general. Accordingly, we assert that, until it is proved otherwise, *all causes may contribute to all effects.*

In more precise terms, we assert as a guide in formulating theories of interaction the ***principle of equipresence:*** *A quantity present as an independent variable in one constitutive equation is so present in all, unless forbidden by some general principle or condition of material symmetry.*

In the older theories deformation was regarded as the "cause" of stress, and temperature gradient as the "cause" of heat flux. Such a separation of causes and effects into categories is unnatural and unjustified by physical principle. Resulting only from the gradual discovery of individual phenomena, it reflects old opinions that break physics up into compartments. Theorists should not propose constitutive equations which artificially divert theories into disjoint channels.

The same view results also from statistical models of continuous matter. There, stress and heat flux appear as gross or mean expressions of a purely mechanical process. The separation of stress as arising from deformation and flux of energy from changes of temperature emerges, not without reason, as a first approximation, but on a finer scale, it is illusory.

Let it not be thought that this principle would invalidate the classical separate theories in the cases for which they are intended, or that no separation of effects remains possible. Quite the reverse: The various general principles, when brought to bear upon a general constitutive equation, have the effect of restricting the manner in which a particular variable, such as the spin tensor or the temperature gradient, may occur. The classical separations may always be expected, in one form or another, for small changes — not as assumptions, but as proven consequences of invariance requirements. The principle of equipresence states, in effect, that no division of phenomena is to be laid down by constitutive equations. It may be regarded as a natural extension of OCKHAM's razor as restated by NEWTON: "We are to admit no more causes of natural things than such as are both true and sufficient to explain their appearances, for nature is simple and affects not the pomp of superfluous causes." This more general approach has the added value of showing in what way the classical separations fail to hold when interactions actually occur.

The "causes" or independent variables of thermomechanics are deformation and temperature:

$$\boldsymbol{x} = \chi(X, t),$$
$$\theta = \underline{\theta}(X, t).$$

The mappings χ and $\underline{\theta}$ assign to the particle X at the time t a place $\boldsymbol{x}$ in Euclidean space and a temperature θ. The ***principle of determinism*** asserts, in general, and that past and present causes determine present effects. If we write

$$f^t(s) \equiv f(t-s), \qquad 0 \leqq s < \infty,$$

then f^t is the *history* of f. In particular, we shall write $\theta^t \equiv \underline{\theta}(X, t-s)$. The principle of determinism asserts that causes operate only through their histories, and the principle of equipresence, that causes are not to be set apart in categories by the theorist. Thus the two principles when applied to thermomechanics are expressed as functional relations giving the *thermomechanical effects* $\boldsymbol{T}, \boldsymbol{h}, \varepsilon, \eta$:

$$\left.\begin{matrix} \boldsymbol{T} \\ \boldsymbol{h} \\ \varepsilon \\ \eta \end{matrix}\right\} \text{ at } X, t \text{ are functionals of } \left\{\begin{matrix} \chi^t \\ \theta^t. \end{matrix}\right.$$

As far as the principle of determinism is concerned, the functionals for the particle X have as their arguments the histories χ^t and θ^t for all particles Z in the body. So as to exclude action at a distance as a mechanism of material response, we impose the ***principle of local action***, demanding that for any given material, two histories which differ from one another only outside some neighborhood of X at the present and all past times shall give rise to the same present effects at X. The four constitutive functionals thus become functionals of χ^t and θ^t in some neighborhood of X.

Finally, the principle of ***material frame-indifference*** is laid down in two parts:

1. Under an arbitrary change of frame, defined by the transformation

$$\boldsymbol{x}^* = \boldsymbol{Q}(t)\,\boldsymbol{x} + \boldsymbol{c}(t),$$

where $\boldsymbol{Q}(t)$ in an orthogonal tensor and $\boldsymbol{c}(t)$ is a vector, the motion χ^* is given by

$$\boldsymbol{x}^* = \chi^*(X, t) \equiv \boldsymbol{Q}\chi + \boldsymbol{c},$$

while the temperature θ and the thermomechanical effects are indifferent:

$$\theta^* = \theta, \quad \boldsymbol{T}^* = \boldsymbol{Q}\boldsymbol{T}\boldsymbol{Q}^T, \quad \boldsymbol{h}^* = \boldsymbol{Q}\boldsymbol{h}, \quad \varepsilon^* = \varepsilon, \quad \eta^* = \eta.$$

2. Under an arbitrary change of frame, all constitutive equations are indifferent.

This principle states, first, that the temperature, the contact force, the contact flux of energy, the internal energy, and the entropy have intrinsic meaning, independent of the observer, and, second, that material properties are the same for all observers.

A material is called *simple* if the thermodynamical effects are determined by the simplest aspects of the histories of the motion and the temperature compatible with the principle of frame-indifference, namely, $\boldsymbol{F}^t$ and θ^t, where

$$\boldsymbol{F} = \nabla \chi_\kappa(\boldsymbol{X}, t) \equiv \nabla \chi\left(\boldsymbol{\kappa}^{-1}(\boldsymbol{X}), t\right)$$

is the gradient of the motion referred to the fixed reference configuration $\boldsymbol{\kappa}$, that is, $\boldsymbol{X} = \boldsymbol{\kappa}(X)$. We include effects of material and thermal inhomogeneity by allowing the response functionals to depend upon $\boldsymbol{X}$ and the present value of grad θ as parameters. Of course they depend also on the choice of the reference configuration $\boldsymbol{\kappa}$. To simplify notation, we do not indicate $\boldsymbol{X}$ and $\boldsymbol{\kappa}$ explicitly, and we set $\boldsymbol{g} \equiv \operatorname{grad} \theta$. The constitutive equations defining a simple material then become

$$\begin{aligned}
\boldsymbol{T} &= \mathfrak{T}(\boldsymbol{F}^t, \theta^t; \boldsymbol{g}) = \boldsymbol{T}^T, \\
\boldsymbol{h} &= \mathfrak{H}(\boldsymbol{F}^t, \theta^t; \boldsymbol{g}), \\
\varepsilon &= \mathfrak{e}(\boldsymbol{F}^t, \theta^t; \boldsymbol{g}), \\
\eta &= \mathfrak{h}(\boldsymbol{F}^t, \theta^t; \boldsymbol{g}).
\end{aligned}$$

The principle of local action is satisfied trivially, while the principle of material frame-indifference leads to functional equations restricting $\mathfrak{T}$, $\mathfrak{H}$, $\mathfrak{e}$, and $\mathfrak{h}$, but I shall not write these.

We have seen already that the principles of conservation of mass and balance of momentum, moment of momentum, and energy effect no restriction on the possible fields of stress, heat flux, and internal energy. Hence they do not restrict the form of the constitutive equations. Following up successful treatments of special cases in earlier work done jointly with Mr. NOLL and Mr. MIZEL, Mr. COLEMAN interprets the Clausius-Duhem inequality as an *identity* which any constitutive equation must satisfy for *all* histories of deformation and temperature. Thus the principles we have stated fall into three categories:

1. Restrictions upon motions and energies when a constitutive equation is given:

a) Conservation of mass
b) Balance of momentum
c) Balance of moment of momentum
d) Balance of energy

2. Rules for forming constitutive equations:
e) Principle of determinism
f) Principle of equipresence

3. Restrictions upon constitutive equations:
g) Principle of local action
h) Principle of material frame-indifference
i) Principle of entropy growth.

We shall consider the last first. Using the equation of energy to eliminate the heat absorption q, we find an equivalent form for the Clausius-Duhem inequality:

$$\varrho(\theta\dot{\eta}-\dot{\varepsilon})+\operatorname{tr}(\boldsymbol{T}\boldsymbol{D})+\frac{1}{\theta}\boldsymbol{h}\cdot\operatorname{grad}\theta\geqq 0.$$

Now there is no term left to adjust. If we substitute for $\boldsymbol{T}, \boldsymbol{h}$, ε, and η the functionals giving them in terms of $\boldsymbol{F}^t$, θ^t, and $\boldsymbol{g}$, we obtain a functional inequality which must be satisfied *identically* in the arbitrary histories $\boldsymbol{F}^t$, θ^t and the arbitrary vector $\boldsymbol{g}$. Such an identical inequality severely restricts the forms possible for the functionals.

Before stating the restrictions, let me describe the idea in more classical terms. In analytical dynamics, a change is said to be "virtual" if the dynamical equations need not be satisfied by it. While we have treated the body force $\boldsymbol{b}$ and the heat absorption q as arbitrarily assignable, in a particular problem in continuum mechanics they will be already given and hence not at our disposal for adjustment so as to balance the linear momentum and the energy corresponding to an arbitrary motion and temperature history. Running the danger of causing confusion by using terms which have for various people various further associations either irrelevant or unnecessary here, I can describe Mr. COLEMAN'S axiom as asserting that *every constitutive equation must satisfy the second law of thermodynamics for all virtual motions and virtual temperature histories.*

The assumptions made so far are not quite sufficient to obtain a thermodynamic theory. If a material is of such a nature that the influence of long past effects remains ever fresh and unforgotten, we cannot expect a trend to equilibrium, and hence we cannot expect to find in thermodynamics any position for classical thermostatics as a theory for processes not too far removed from equilibrium. Let $\boldsymbol{\Delta}^t$ stand for the deviation of the deformation-temperature history $\boldsymbol{F}^t$, θ^t from the present deformation and temperature:

$$\boldsymbol{\Delta}^t(s)=\left(\boldsymbol{F}^t(s)-\boldsymbol{F}(t),\ \theta^t(s)-\theta(t)\right).$$

The magnitude of this deviation is then

$$|\boldsymbol{\Delta}^t(s)|=\sqrt{\operatorname{tr}(\boldsymbol{F}^t-\boldsymbol{F})(\boldsymbol{F}^t-\boldsymbol{F})^T+(\theta^t-\theta)^2}.$$

We can then define the *recollection* of the deformation-temperature history as

$$\|\boldsymbol{\Lambda}^t\|_h = \sqrt{\int_0^\infty |\boldsymbol{\Lambda}^t|^2 (h(s))^2 ds},$$

where $h(s)$ is a positive real function which vanishes sufficiently rapidly as $s \to \infty$. $h(s)$ is called an *obliviator*, since $1/h(s)$ is the density of oblivion allotted to the time s units before the present time in calculating the recollection $\|\boldsymbol{\Lambda}^t(s)\|_h$. Thus large values of $\boldsymbol{\Lambda}^t(s)$ at sufficiently long past times $t-s$ contribute less to the recollection $\|\boldsymbol{\Lambda}^t\|_h$ than do equal values at recent times. Since $\|\boldsymbol{\Lambda}^t\|_h$ is a norm, it defines a topology in the function space of deformation-temperature histories, and a neighborhood of the rest history $\boldsymbol{\Lambda}^t(s) \equiv \mathbf{0}$ in this topology consists of those deformation-temperature histories which are near to the rest history in the sense of recollection. That is, any large deviations from rest have occurred long ago.

COLEMAN and NOLL have given a precise sense to the familiar idea that a material should respond less to causes operating long ago than to recent ones by demanding that the response functionals be smooth in the topology defined by recollection. Continuity, for example, means that effects — stress, heat flux, energy, entropy — corresponding to a deformation-temperature history of small recollection shall be very nearly those corresponding to rest. Roughly speaking, a material has fading memory if and only if it exhibits relaxation of stress, heat flux, energy, and entropy. The thermodynamic theory of COLEMAN rests on a stronger *principle of fading memory*. Namely, let $\mathfrak{F}$ be any one of the four response functionals, and write it in the form

$$\mathfrak{F}(\boldsymbol{F}^t, \theta^t; \boldsymbol{g}) = \mathfrak{F}(\boldsymbol{\Lambda}^t; \boldsymbol{F}, \theta, \boldsymbol{g}).$$

Then COLEMAN assumes that for each fixed $\boldsymbol{F}, \theta, \boldsymbol{g}$, the response functional $\mathfrak{F}$ is twice Fréchet-differentiable in a neighborhood of the rest history, $\boldsymbol{\Lambda}^t \equiv \mathbf{0}$. This assumption, added to the preceding ones, makes it possible to derive a true thermodynamics of deformation.

COLEMAN'S fundamental theorem consists in four assertions, which I shall present and explain in turn. First, $\boldsymbol{g}$ drops out of the functionals $\mathfrak{e}$ and $\mathfrak{h}$. That is, the internal energy ε and the entropy η are unaffected by the temperature gradient. In particular, the free energy is a functional of the deformation-temperature history alone:

$$\psi \equiv \varepsilon - \eta\theta = \mathfrak{p}(\boldsymbol{\Lambda}^t; \boldsymbol{F}, \theta).$$

Second, the stress and entropy functionals are determined from the free-energy functional by a kind of functional differentiation. Let $D_{\boldsymbol{F}}$ and D_θ denote the operations of differentiation with respect to the present deformation and temperature, respectively, when the *past* histories of both these fields are held constant. Then the second part of COLEMAN'S theorem asserts that the functionals $\mathfrak{T}$ and $\mathfrak{h}$, which give the stress and the entropy, are determined from the free-energy functional $\mathfrak{p}$ as follows:

$$\boldsymbol{T} = \mathfrak{T}(\boldsymbol{F}^t, \theta^t) = \varrho \boldsymbol{F}[D_{\boldsymbol{F}} \mathfrak{p}(\boldsymbol{F}^t, \theta^t)]^T,$$
$$\eta = \mathfrak{h}(\boldsymbol{F}^t, \theta^t) = -D_\theta \mathfrak{p}(\boldsymbol{F}^t, \theta^t).$$

In the special case when $\mathfrak{p}$ reduces to a function of $\boldsymbol{F}$ and θ, these relations assume the forms familiar in classical thermostatics. More generally, there is no caloric equation of state, and the free energy is affected by past deformations and temperature changes; nevertheless, COLEMAN'S result shows that the free-energy functional determines the stress functional and the entropy functional.

Let $\delta \mathfrak{F}(\ldots | \boldsymbol{\Gamma})$ stand for the first Fréchet differential of $\mathfrak{F}$ corresponding to the increment $\boldsymbol{\Gamma}$, and set

$$\sigma \equiv -\frac{1}{\varrho} \delta \mathfrak{p}(\Delta t; \boldsymbol{F}, \theta | \dot{\boldsymbol{F}}^t, \dot{\theta}^t).$$

Then $-\sigma\theta$ is the rate of increase of free energy when $\dot{\boldsymbol{F}}$ and $\dot{\theta}$ are suddenly set equal to zero, while the past histories of deformation and temperature are held constant. Specifically,

$$\dot{\psi} = \frac{1}{\varrho} \operatorname{tr}(\boldsymbol{T}\boldsymbol{D}) - \eta\dot{\theta} - \theta\sigma.$$

In classical thermodynamic formulae, $\sigma = 0$. More generally, σ may be called the *internal dissipation*. The third part of COLEMAN'S theorem asserts that

$$\sigma \geqq 0.$$

While the specific entropy itself may decrease or increase according to circumstances, the internal dissipation is never negative.

The fourth part of COLEMAN'S theorem asserts that

$$\boldsymbol{h} \cdot \operatorname{grad} \theta \geqq -\varrho\theta^2\sigma.$$

In classical theories of dissipation, heat flows against a temperature gradient: $\boldsymbol{h} \cdot \operatorname{grad} \theta \geqq 0$. We recover this same result when $\sigma = 0$. In more general cases, when $\sigma > 0$, the extent to which $\boldsymbol{h}$ and grad θ may fail to subtend an acute angle is limited by the internal dissipation occurring and the temperature at which it occurs.

Since this seminar concerns the philosophy of science, I have neither stated the assumptions in mathematical precision nor even outlined the logical argument. Also, I shall not present any of the further developments of the theory, though some need to be mentioned so as to make it abundantly plain that success has been achieved not only in unifying old ideas but also in discovering important new results. First, every material included by COLEMAN'S thermodynamics has a caloric equation of state approximately in two special kinds of circumstances, and, in general, only for these, namely, in the limits of infinitely slow or infinitely fast deformation. Second, the purely static minimal principles of GIBBS are given a status as assertions about processes with given, fixed $\boldsymbol{F}$ and ψ or ε or η. Third, the inequalities of PLANCK, which heretofore had seemed to be either new postulates or else assertions about some kind of virtual change of systems in equilibrium, come out as proved theorems, valid in all processes. Fourth, the work theorem for cyclic processes which are isothermal, iso-energetic, or isentropic, appears as a proved result, valid in all processes. Fifth, every material can transmit acceleration waves, and the laws of propagation of these waves are just the same as those of an associated hyperelastic material, the stored-energy function of which depends upon the history of deformation and temperature.

Thus far, no special symmetries have been attributed to the material. Every material has its own group of symmetries, and these may differ according to the local reference used. Indeed, if a unimodular tensor $\boldsymbol{H}$ is such as to satisfy the two equations

$$\mathfrak{p}(\boldsymbol{F}^t, \theta^t) = \mathfrak{p}(\boldsymbol{F}^t\boldsymbol{H}, \theta^t),$$

$$\mathfrak{h}(\boldsymbol{F}^t, \theta^t; \boldsymbol{g}) = \mathfrak{h}(\boldsymbol{F}^t\boldsymbol{H}, \theta^t; \boldsymbol{g}),$$

identically in $\boldsymbol{F}^t$, θ^t, and $\boldsymbol{g}$, then $\boldsymbol{H}^{-1}$ maps the given local configuration onto another one having the same density and yielding exactly the same response to effects of deformation and temperature. Such a change of configuration, then, is *indistinguishable by experiment* for the material in question. The set of all such $\boldsymbol{H}$ forms a group, which is called the *isotropy group* $\mathscr{g}$ of the material at the given particle and for the reference considered. Most of the physical properties of a material particle are specified in terms of its isotropy groups. A material is

$$\left.\begin{array}{l}\text{a fluid}\\ \text{a solid}\\ \text{a subfluid}\\ \text{isotropic}\end{array}\right\}\ \text{if there exists a reference such that}\ \left\{\begin{array}{l}\mathscr{g} = \mathscr{u},\\ \mathscr{g} \subset \mathscr{o},\\ \mathscr{g} \supset \mathscr{h},\\ \mathscr{g} \supset \mathscr{o},\end{array}\right.$$

where $\mathscr{u}$ is the unimodular group, $\mathscr{o}$ is the full orthogonal group, and $\mathscr{h}$ is a dilatation group

I do not think it possible to make concrete and correct observations on the philosophy of science without regard for what science is. Mechanics and thermodynamics have had their influences upon philosophy in the past, but the sciences themselves have changed meanwhile, and I see few signs that philosophical circles have made themselves aware of anything since the extreme formalism of LAGRANGE and HAMILTON, the revival of schoolboy mechanics by MACH, and the murky gloom of Victorian steam engines and reversible processes in a stagnant universe. We must sympathize with the philosopher, since he is likely to expect to get a true picture of classical mechanics and thermodynamics by looking in a physics book. Some of the older formalism, and a smaller part of the old conceptual structure, remain valid today as special cases, often degenerate ones, but the whole view of what mechanics is for and how it should be done has changed.

First, mechanics and thermodynamics have merged into a single science. Second, the emphasis in this science lies on deformation rather than inertia; if you prefer, on configuration rather than mass. The old mechanics concerned, in the main, singularities of inertia, filling a nowhere dense set in space; the old thermostatics concerned dense uniform blocks. The merger of concepts was made possible by the shift of emphasis from curves to general functional transformations, which is a normal aspect of modern mathematical thought. In the period between the two wars, the program of LEIBNIZ was neglected: "after consideration of these few things, this whole matter is reduced to pure mathematics, which is the one thing to be desired in mechanics and physics." While physicists grew to regard mathematics like a sausage grinder, the mathematicians turned away from physical problems like garbage. Both forgot that western mathematics and western physics grew up together as partners, each with a degree of independence and a personality of his own, but each helping the other. Modern natural philosophy returns to the old program of making *the physical concepts themselves* mathematical from the outset, and mathematics is used to *formulate* theories.

The usual dividing line for mathematics stops just after geometry, cutting off mechanics. "Space", "point", "line", "figure", "incident", "parallel", "volume", and "area" are mathematical terms, having properties embodied in specific, unequivocal axioms, with no mention of experiment or observation. "Mass" and "force" belong to physics. No mathematical axioms are laid down for them, and much is said about how to measure and interpret them. Specific theories of the action of forces on masses are set up by drawing pictures or repeating rituals, and the mathematics does not come until afterward. It is not a question of rigor: When finally the mathematics gets started, it may be entirely precise, but setting up the theory is an extra-mathematical operation. Sometimes it is said that while geometry is a part of mathematics, mechanics and the rest of physics are experimental sciences.

Although this statement may describe current practise, it is not a divine order. There is also an experimental geometry. At one time, it was doubtless quite a problem to calculate areas and volumes, and for some odd shapes experimental methods are used even today. However important this practical side may be, universities no longer recognize it among their manifold community services, for mathematical geometry has been so widely accepted that we are led to believe we understand all of geometry when we have mastered the mathematics of it. No laboratories are provided where the student may test his knowledge of areas by cutting out and weighing sheets of tin with curved boundaries. Indeed, even the most practical of partical men is likely to have a rather inverted picture of geometry. When the carpenter uses a little more wood than planned, he does not say, "The theory gave the wrong answer, since it neglects some of the real physics," but "The boards weren't square." Rather than trying to fit theory to practice, he reproaches application for failing to reach the precision of theory!

Rational mechanics is an extension of geometry, with similar aims, methods, and results. Both are sciences of experience. Neither is experimental. Experimenters and appliers, be they physicists, engineers, or practical men, make good and successful use of both sciences. It has been so many centuries since geometry was distilled from its experiential basis that no one doubts its exactness or asks how or when to apply it. I have never heard of a geometry teacher's being troubled by an engineer who says, "Look at this beam and tell me whether its cross-section is a triangle or a circle!" The practical man has his own methods of setting theory into correspondence with experience; indeed, such correspondence furnishes his specific business; he does not expect geometry itself, or the geometer, to tell him *when* to use $A=d^2$ instead of $A=\pi r^2$ to find the area of a particular table top. Neither does he discourage the geometer with an accusation like those thrown commonly upon the natural philosopher today, "Your theory represents a triangle as having smooth, continuous edges, and as being of no thickness, but any real triangle is made up of molecules, its edges are all ragged with holes and gaps, and it has at best some approximate or average thickness. How can your geometry be of any use for physics if it neglects these fundamental facts about the structure of matter?" The practical man neither despises geometry as useless nor worships it as an imperial pill to purge all his troubles and render his function null.

Why is it that the physicist who is content to learn geometry in all precision reverses his stand the moment the concepts of mass and force are added to space and distance, crying that when "physics" enters the picture, rigor must be kicked out by "intuition", and wholly different methods and criteria are in order? He is failing to see that while physics through the centuries has taken over many domains formerly thought parts of philosophy, as the second stage mathematics takes over domains formerly thought parts of physics. Geometry was consolidated as mathematics two thousand years ago; conquest of mechanics for mathematics began three hundred years ago; and is now in progress, against resistance. Indeed, often physical scientists, instead of learning and using modern mechanics as they do geometry, with gratitude for what it is, decry it for what it is not, either reproaching the theorist for "neglecting the physics" or suggesting it would be better if he "stuck to mathematics", by which they mean infinitesimal extrapolations of the mathematics they themselves half learned in school. Rational mechanics *is* mathematics, just as geometry is mathematics.

The generality and abstraction of the theories of COLEMAN and NOLL are only common and natural in the course of mathematics. To construct the real number system requires a more abstract level of thought than do operations with rational numbers; all experimental data is reported in terms of rational numbers alone; yet for the past two hundred years even grammar school children have been taught to operate with real numbers, because it is simpler to do so, and the disadvantages of an arithmetic excluding irrationals are too obvious to need detailing. It is the process of generality and abstraction that reduced analytical mechanics, some two hundred years ago, to a form which can be taught to university students today in one semester. The advantages of the new and abstract approach to materials are so clear as to want no propaganda, and its victory is certain. The resistance of the old guard is that which has greeted every advance in mathematical science and always dies away with one generation.

The ingratitude of applied scientists for the fundamental work they themselves appropriate, often after a period of rejecting it as abstract and "useless", and later turn to profit in their more lucrative professions, sometimes grows out of failure to understand what it really is they have digested. Most natural scientists have a strong bias toward experiment; all of us have been subject to footnotes, books, or even courses in the history of science that distort the growth of scientific thought in the past as wickedly as a communist text distorts political history.

The hard facts of classical mechanics taught to undergraduates today are, in their present forms, creations of JAMES and JOHN BERNOULLI, EULER, LAGRANGE, and CAUCHY, men who never touched a piece of apparatus, and their only researches that have been discarded and forgotten are those where they tried to fit theory to experimental data. They did not disregard experiment; the parts of their work that are immortal lie in domains where experience, experimental or more common, was at hand, already partly understood through various special theories, and they abstracted and organized it and them.

I am not suggesting that no experiments should be done; I am not even saying that experimenters are doing too many experiments. I am merely pointing out that dabbling in data is no way to make a great theory. The role of theory is not to adjust fudge factors to data, but, like geometry, to draw much from little. Herein lies its economy, its beauty, and, ultimately, its use. EULER in deriving the laws of motion of a rigid body remarked that they would be just as valuable if there were no rigid bodies in the physical world. Once the general ideas of mechanics are understood, certain particular materials, such as the rigid body, the perfect fluid of Euler, and the simple fluid of Noll, present themselves as naturally as do triangles and circles in geometry, and it is a mathematical necessity to study them. Applications are neither sought nor despised: they come of themselves. Classical geometry is so obviously, so egregiously applicable that no one asks what its applications are. Applications raise the human value of geometry and mechanics but do not change their natures.

The analogy to geometry should be drawn further. The early geometry laid out much thought on certain particular figures; triangle, square, circle, etc. Higher geometry requires a different method. It would be simply wasteful to go on in the same way with figure after figure. Rather, in modern geometry we learn properties of infinite classes of figures, such as polygons, conic sections, algebraic or analytic curves, and then take up more inclusive ideas such as metric, connection, and manifold. The geometric concepts are expressed directly in terms of *properties of invariance.*

So it is in rational mechanics. Not only to spare time but also to think simply and surely, we approach problems of material response in generality. Our physical experience with materials is summed up in statements of invariance. In the work of Noll and Coleman, these are put directly into *mathematical language.* The theorist can no more be expected to tell the experimenter what constitutive functional to use than Euclid tells the carpenter when to use a circle or when to use a square, or, in the terms preferred by physicists, "states the range of applicability and estimates the experimental error" of a triangle.

There are also persons who dismiss the modern work as being "axiomatics". While, unfortunately, we have not yet reached the point where a fully axiomatic treatment seems to be in order, the implication that things everyone knew already are just done over in a rigorous way is calumny. Not only the viewpoint but also some of the specific theorems I have summarized are new, some not yet published. Some old results are embedded in the new structure; those that are right deserve to be embedded, and a new theory not including them would be a wrong theory. The modern work is *mathematical* and hence *strictly logical,* like classical geometry. Recall that after attempts scattered over three millenia and three continents, a proper set of axioms for Euclidean geometry was first obtained by Hilbert, in 1899. The older geometry, while not successfully axiomatic, was also not illogical, since it employed only logical methods. Rational mechanics, a logical or mathematical science, is in a pre-Hilbertian stage, and a completely axiomatic treatment lies some years ahead, but a start has been made.

Chapter 5

Levels of Description in Statistical Mechanics and Thermodynamics[1]

Harold Grad

Courant Institute of Mathematical Sciences
New York University

Dedicated to the memory of
Bernard Friedman
whose life was a source of inspiration both as a person and as a mathematician.

1. Introduction

The motivation for this talk is the belief that in any study of the philosophy of science a necessary prerequisite is that the input data, which is science itself, should be correct. In statistical mechanics particularly, precision is an elusive goal. It is safe to say that a major portion of the nontrivial results in statistical mechanics has been derived from inconsistent formulations. Now, as we know, *all conclusions* follow logically from inconsistent postulates. This is indeed observed in practice. Fortunately there is a natural filter which shields from publication the most whimsical, outlandish, and flagrantly incorrect conclusions. But occasionally such a result is enshrined as a *paradox*, and then heroic efforts are required to dislodge it.

Of course, mathematical rigor alone is not the solution to these difficulties. It is possible to join unimpeachable mathematical analysis to a faulty or irrelevant formulation of the physical problem. We shall try to use the best available tools (when necessary predicting what form we expect they will take!) on a specific problem, the prediction of observable phenomena from the laws of dynamics.

Many areas of mathematical physics are blessed with an abundance of alternative ways of describing the same physical phenomena. One can describe a gas in terms of the motions of its individual molecules, or in terms of a smooth molecular distribution function as in the Boltzmann equation, or by a variety of macroscopic fluid coordinates. Insofar as they all describe the same physical problem, they must be in some

[1] This work was supported by the Air Force Office of Scientific Research under Grant No. AF-AFOSR-815-66.

sense similar. But the differences and even basic contradictions lie much closer to the surface than the similarities. If only a molecular description is complete, how can the much cruder fluid descriptions ever be found to be accurate? How can the reversible particle orbits be compatible with the irreversible Boltzmann equation and with a choice of both reversible and irreversible fluid formulations? Must we invoke Maxwell demons to perturb the ordinary differential equations which govern molecular motion to justify the empirical success of purely stochastic hypotheses which entail randomness? Or are these stochastic crutches a consequence of the exact dynamical equations of motion which are powerless to resist a probabilistic demand that they docilely follow a "most likely" path? And finally, are analysts only holding a rear guard action until computing machines become large enough and fast enough to obtain all results from first principles?

The basic question which we raise is whether all observable properties of matter are, in principle, deducible from the laws of motion which govern individual particles[1] so that a sufficiently competent (and wealthy) mathematician could deduce all observable phenomena and reliably predict the outcome of an experiment from his armchair (or timesharing console). The simplest (naive) answer is "of course"; and as an immediate (naive) consequence, all lower level descriptions (such as thermodynamics) are incomplete and inherently approximate. In practice this answer is violated in both directions. On the one hand there are many empirical examples where a very much lower order of information appears to be essentially exact: thermodynamics, fluid dynamics, transport theory, etc. There are even a few cases where they can be proved to be exact. On the other hand, the detailed particle description is frequently treated as though it were an incomplete formulation of the laws of nature, requiring supplementation by auxiliary laws, frequently stochastic: *a priori* probability proportional to volume in phase space, molecular chaos, observable states selected by maximum entropy, high order correlations dependent on lower order, etc.

A more careful examination of this question shows that the dual difficulties that the equations of motion are too detailed for practical use and at the same time seem to be incomplete are two aspects of the same phenomenon. An illustration is the rarefied gas. At the same time (i.e., in a certain limit) that prediction of the motion of an individual particle becomes impossibly sensitive to slight changes of its initial state, other variables emerge which are increasingly insensitive to these "improper" variables and take over determinism. It would be extremely repugnant to be forced to accept the necessity (as distinguished from

[1] We consider only classical mechanics, but much of the discussion should apply qualitatively to quantum systems.

convenience) of supplementary laws of physics to describe large systems when the laws for small systems are felt to be complete. Our position is that the accessory (usually stochastic) hypotheses are logical consequences of the laws of motion and are therefore logically redundant. In mathematical terms, some properties of certain systems of very many differential equations approximate random behavior. The qualitative behavior of large systems can be quite different from that of small systems. Only some theorems of this type can be proved at present. But enough is known to remove any lingering doubts. A practical difficulty is that a stochastic statement can be a very good approximation while being strictly incompatible with the laws of motion. It is unfortunate that there does not exist at present a logical apparatus for dealing with "slightly" incompatible logical systems.

Our program is to demonstrate that the various redundant hypotheses which are introduced into statistical mechanics to supplement the equations of motion are unnecessary; and to examine the various levels lower than a full dynamical description to distinguish those which become exact in some limit from those which are *ad hoc* and provisional empirical approximations (such as a truncated power series or Fourier series). For example, the Boltzmann equation becomes exact in a certain well-defined limit (a rarefied gas). The equations of gas dynamics become exact in another limit, performed on solutions of the Boltzmann equation. But moment approximations intercalated between the Boltzmann description and fluid description are approximations which can be successively improved by taking more terms. Similarly, there seems to be no exact kinetic equation comparable to the Boltzmann equation for a dense gas or liquid[1].

We maintain that it is unnecessary to introduce random perturbations by the external world to verify irreversibility, approach to equilibrium, and other seemingly stochastic effects[2]; it cannot be necessary to invoke principles such as maximum entropy (or information) to obtain rigorous transport coefficients; it must be possible to justify from first principles functional dependence of higher order correlations on lower whenever it happens to be truly justified (as it is for an ideal gas but is not for a dense gas).

Practically speaking, these expedients are exceedingly useful and possibly even inevitable in the present state of the art. The wholesale (and even indiscriminate) use of probability is what makes the difficult and sophisticated study of statistical mechanics accessible to undergraduates. The great success of mathematical physics lies in its frag-

[1] This was pointed out by GRAD [*1*] and has been more firmly established by recent detailed calculations [*2*—*7*].

[2] For an eloquent contrary opinion see KAC [*8*] and also MAYER [*9*].

mentation. There is a multitude of models, each one giving satisfactory results within a certain restricted domain. The science of physics is partly rational and partly empirical witchcraft. In science as in business or in politics, success provides its own justification. Our question is a basic (and frequently impractical) one; what is the structure of the whole? In the course of this lecture we should be forgiven if we dismiss many very important and elegant theories simply because they do not bear on this question.

Although all information does reside within a molecular description, the manner in which it can be extracted is exceedingly subtle. We can demonstrate this by a simple estimate of the size of a representative numerical computation. Let us consider a macroscopic problem which one might eventually hope to solve numerically by computing particle orbits. The ordinary test of reliability would be to reverse the velocities at the end of the computation and recover the initial state within some specified accuracy. In air under normal conditions, a concentration gradient in a one meter tube of one centimeter bore could take about 10^4 seconds to equilibrate. During this time a given molecule would collide about 10^{14} times. Since the diameter of a molecule is about one percent of the mean free path, an error in the initial position or velocity of the molecule would be magnified by a factor $(100)^n$, $n=10^{14}$ by the end of the computation. Disregarding the computation itself, the initial storage of coordinates for 3×10^{21} particles would be at least 10^{37} bits. According to present engineering practice this would require a memory storage unit on the order of 10^7 times the volume of the earth. The time for the computation would enormously exceed the present age of the universe.

The first and most obvious question is, how does the gas itself (considered as an analogue computer) solve this mathematical problem? The answer is that it does not. An error of 10^{-n} cm in position or 10^{-n} cm/sec in velocity of a single molecule $(n=10^{14})$ would destroy the accuracy of the computation; this is approximately 10^{2n} smaller than the quantum uncertainty. In this sense, uncertainty is just as basic to classical as to quantum mechanics.

We can only conclude that nature would be completely chaotic and unpredictable in large systems were it not for a peculiar insensitivity to certain types of perturbations; we fully expect differential equations to be similarly insensitive. It is a trivial point that no macroscopic effect is noticed at a given instant if one molecule is displaced by a small amount. What is not at all trivial is that an astronomical number of collisions after such a small perturbation, when one finds most of the molecules displaced by appreciable amounts and some of them by very large macroscopic amounts, there are some properties of the gas which are almost unaffected. This is not even necessarily true. In a macroscopically unstable (say turbulent) regime, it is conceivable that a microscopic perturbation could exert a finite macroscopic effect at some later time. In this case not only the microscopic state but the naive macroscopic state is not predictable, and a still coarser level of description must be sought in order to restore determinism.

A numerical computation of orbits could be attempted if one gave up the goal of accurately predicting individual orbits. If there really are other, less sensitive, coordinates of the gas, they may be insensitive to the burgeoning numerical error. But this procedure loses the self-contained immaculate status of the direct numerical computation. The calculation becomes a numerical experiment whose laws are to be discovered empirically and which *may* turn out to be an analogue of the physical experiment.

If it is true that some properties of large dynamical systems are insensitive to certain perturbation of the initial data, then this is an important mathematical property of large systems of differential equations which should be mathematically verifiable. But such theorems require more sophisticated tools than the ones which lead to the conventional existence theorems and estimates of the growth of solutions.

We can point out two divergent possibilities in a system of many degrees of freedom. The equations may be exceedingly stable so that a cumulation of small errors will not result in catastrophe. Or the system may become exceedingly sensitive to small perturbations. The most interesting feature of a large dynamical system which represents a gas in a box is that it exhibits both trends simultaneously. As one set of variables becomes increasingly difficult to follow, another qualitatively different set becomes more and more insensitive to small changes. On the other hand, we have systems of even *infinite* degrees of freedom, represented by partial differential equations, which are very well behaved. Their approximation by finite differences can be interpreted as a microscopic model. Thus there is no universal behavior which can be expected from all very large systems.

The single feature which distinguishes statistical mechanics from ordinary mechanics is the large number of degrees of freedom. Under certain circumstances the fact that N is large dominates the scene. The fact that the system is governed by specific differential equations is secondary, and the laws of dynamics are almost powerless to overcome the overwhelming consequences of large N. To a certain degree of approximation, the behavior of the completely deterministic system of equations appears to be random. On the other hand, this is not always so. We shall give elementary examples in which the dynamical laws restrict the motions to "extremely unlikely" situations. The correct result is given by a delicate balance between dynamics and large N. Thus the use of statistical methods, essentially ignoring the dynamics, is a very powerful tool; but like a very powerful drug, it must be used with proper safeguards and under adult supervision.

The simplest example of the use of probability is given by the study of equilibrium. With overwhelming probability, a state chosen at random will be approximately uniform in density and Maxwellian in velocity. Taken at face value, this implies that there is negligible liklihood that one will ever encounter a state which is not in equilibrium! Thus all fluid dynamics is concerned with extremely improbable events (from which even less probabable events must be carefully excluded). Even in equilibrium, finite angular momentum can be carelessly discarded as being extremely unlikely. We must exercise great care in dismissing sets of small measure (low probability) lest we throw out the baby with the bath water. And the use of plausible statistical estimates and hypotheses under such circumstances is evidently a matter of great delicacy. A stochastic hypothesis must always be considered to be tentative because of the possibility that the dynamics will assert itself and deny the primacy of the "most probable" behavior. One finds disarmingly simple proofs of extraordinarily subtle phenomena in statistical mechanics. But examination frequently shows that the same "proof" gives incorrect answers in other, equally plausible cases. The correct answer is sometimes the most probable state. But this is not a postulate; it is (when true) a derivable consequence of the laws of motion.

In summary, the empirical observation that different mathematical formulations should be expected to give accurate descriptions of the same physical situation suggests that the relation between the two formulations can be uncovered by purely mathematical investigations. Such a connection can always be expected to be subtle and mathematically singular. A well-understood example in fluid dynamics is the relation between inviscid and slightly viscous flow. One must beware of boundary layers and D'ALEMBERT'S paradox. The differences in the models which are compared become much more pronounced in statistical mechanics where there is an astronomical range in the choice of a description. An unsophisticated attitude finds paradoxes in the fact that two qualitatively different approaches can yield quantitatively indistinguishable answers. The facts that reversible particle dynamics can approximate arbitrarily closely to the irreversible Boltzmann description and that a simple thermodynamic or fluid description can approximate arbitrarily closely to a detailed molecular description represent very difficult, singular, and even subtle mathematical results. But they are paradoxes only in the sense that any nonuniform limit is a paradox to the mathematically uninitiated.

Our attitude will necessarily be that of an advocate (if not an evangelist). One reason is that only some of the required results have been proved; (but we must be careful to distinguish a legitimate scientific conservatism in the absence of absolute proof from simple obstinacy). Another reason is that a century of indoctrination with irrelevant paradoxes must be swept away before the real difficulties can be appreciated.

2. Levels of Description

There are very many alternative models ranging from a complete microscopic description of individual molecules to a fully macroscopic fluid analysis which employs a small number of field variables and the thermodynamic description of equilibrium which requires only a small number of absolute constants to completely specify the state. We shall point out only some of the more interesting relationships.

The molecular specification of a state can be described as a point P in a $6N$-dimensional space (for simplicity we take N point molecules, each located by 3 coordinates and 3 velocities). The motion of the system is represented by a curve $P(t)$. Any physical observable can be considered to be known when P is given; it is therefore represented by a specific function $\varphi(P)$. The variation in time of this observable will be implict in the variation of the argument $P(t)$.

With an eye to a possible reduction in irrelevant information, we consider repeating an experiment in which the initial value of P is not precisely repeated but is only partially constrained. To describe this situation, it is conventional to introduce an *a priori* probability density $f_N(P)$. As a consequence of the equations of motion, $f_N(P, t)$ will satisfy LIOUVILLE's equation. Instead of the precise value of an observable $\varphi(P)$ we can only compute certain statistical properties. The most elementary is the expected value of φ, $\langle\varphi\rangle = \int f_N \varphi \, dP$. More information is given by the entire distribution function of the values taken by φ at a given instant. Still more information is available in terms of correlations between the values of φ at two times, or correlations between two observables φ_1 and φ_2, etc. All this information can be computed from $f_N(P, t)$.

Since LIOUVILLE's equation is linear, the basic theory is the same for a singular initial distribution $f_N(P)$ concentrated at a single point (a δ-function). The corresponding solution of LIOUVILLE's equation is identical to the trajectory $P(t)$ with the given initial value. It would appear that the original problem has not been altered by the introduction of probability. But there is a significant qualitative difference between smooth solutions and singular solutions of LIOUVILLE's equation. For example, any expected value, probability distribution, or correlation function for an observable $\varphi(P)$ can be expected to approach a definite limit as $t \to \infty$ for a smooth solution $f_N(P, t)$ but not for a singular solution. Thus we must consider that introduction of a smooth (absolutely continuous) probability density significantly alters the nature of the description, even though both involve comparable amounts of microscopic detail.

To reduce the amount of information we can look for more primitive descriptions in terms of the one-particle distribution $f_1(P_1)$, two-particle distribution $f_2(P_1, P_2)$, etc., where P_1 represents the six coordinates of a single molecule and P represents the collection $(P_1, P_2, \ldots, P_N)$.

The one-particle distribution f_1 is frequently taken to satisfy the Boltzmann equation. The enormous reduction from a function f_N of $6N$ arguments to a function f_1 of 6 arguments becomes exact in a certain well-defined limit. Briefly, we let N become large and at the same time decrease the mass, m, of a molecule and its diameter, σ (more correctly σ is a scaling parameter in the intermolecular potential) such that Nm and $N\sigma^2$ remain finite. In this limit the mean free path is fixed but the total volume occupied by the molecules, $N\sigma^3$, vanishes. The gas becomes ideal in the limit; one aspect of the intermolecular force is lost in that there is no contribution from the potential energy to the equation of state, and all transport is purely kinetic. But collisions remain as a finite mechanism for the time evolution of f_1. This limit removes the graininess of the gas and also reduces fluctuations about the expected value of f_1. In the limit we obtain an ideal gas which satisfies the Boltzmann equation exactly[1]. Continuous dependence on initial data is lost for an individual molecule and for f_N but is transferred to f_1 in this limit. An important consideration is that the scale of the spatial dependence of f_1 is held fixed in this limit.

Another limiting process is the one introduced by BOGOLUBOV [*10*] which yields an unconventional Boltzmann equation in which the arguments of two distributions $f_1(P_1)$, $f_1(P_2)$ of colliding molecules P_1, P_2 are taken at different spatial locations. This distinction would drop out in the scaling adopted above since two colliding molecules are separated by a distance $\sigma \to 0$. But Bogolubov tacitly assumes that the space scaling of f_1 is tied to σ rather than to the mean free path. Thus BOGOLUBOV's equation is not at all the Boltzmann equation; it describes very small scale disturbances on the order of the molecular diameter rather than on the order of the mean free path. Such a disturbance will decay in a time much less than a mean collision time. Thus the collision term

[1] The most careful mathematical presentation of this problem is given in [*1*]. There are a large number of more complicated but less reliable derivations of the Boltzmann equation by purely formal expansions in a parameter or as Fourier series without error estimates.

in BOGOLUBOV'S equation describes a small perturbation to the motion of the gas (which has the same order as the imperfect gas correction).

An estimate of the fluctuations about the expected value of f_1 shows that they are large compared to the correction which BOGOLUBOV'S equation makes over the Boltzmann equation. The fluctuations and BOGOLUBOV'S correction are comparable (and small) if the scale of f_1 is comparable to the mean free path; but the fluctuations dominate for smaller scale disturbances. Thus, although BOGOLUBOV'S equation contains a "correction" to the Boltzmann collision term, the range of validity of the equation is drastically reduced.

There have been many attempts to fill the void between the one-particle Boltzmann description and the N-particle Liouville description. It has become increasingly clear that the gap cannot be filled[1]. A kinetic description in which the mean free path is comparable to the scale of spatial variation can be formulated in terms of f_1 only for the ideal gas using the conventional Boltzmann equation. There is a sequence of *ad hoc* approximations which may give increasing accuracy as one includes f_2, f_3, etc., and digests more and more complicated initial and boundary data. But none of these equations appears to be exact in any limit (as is the Boltzmann equation) except possibly a limit in which the mean free path becomes small compared to the macroscopic scale. But in this limit one obtains only fluid dynamics (i.e., transport coefficients), not a kinetic equation.

A fluid description can be obtained from the Boltzmann equation or, without this intermediate step, directly from particle dynamics. The latter problem is relatively neglected[2].

The Boltzmann equation can be formulated to contain a parameter, ε, which represents the scale of the mean free path. Small ε represents a singular limit, but it has been shown rigorously that every solution f_1 approaches a local Maxwellian as $\varepsilon \to 0$, and the parameters in the local Maxwellian (which are exactly the local fluid state) satisfy the inviscid Euler equations with the same initial fluid state as f_1 [*12*]. More precisely, there is an initial transient of duration ε during which the given f_1 swiftly transforms to within an error ε of a local Maxwellian, after which fluid dynamics takes over. The process by which the detailed microscopic information which is present in f_1 becomes rapidly irrelevant can be precisely examined.

A power series expansion in the small parameter ε gives rise to either the Hilbert or Chapman-Enskog theory depending on exactly how one proceeds. Either series can be shown to be asymptotic to true solutions

---

[1] See footnote 1, p. 51.

[2] A very powerful attack on this problem has been made by MORREY [*11*], but gaps still remain.

of the Boltzmann equation which are sufficiently smooth. Although the variables that enter into the Hilbert and Chapman-Enskog theories appear to be purely macroscopic, the correct fluid initial state that must be supplied depends on the entire microscopic initial state f_1. To lowest order (inviscid Euler equations) only the initial fluid state enters. At each succeeding order in ε additional information is extracted from the initial f_1 and applied to the subsequent fluid behavior. Thus the loss of memory of the detailed initial state is a complex process; it is converted into a perturbation of the fluid state rather than being strictly lost.

There is an alternative method of interpolation between the BOLTZMANN and macroscopic descriptions by approximating the distribution function through polynomials or other convenient functions [*13*]. This is an *ad hoc* procedure whose accuracy is hard to estimate. But it is not so strictly bound to the macroscopic fluid limit since the state variables and differential equations themselves are non-fluid, whereas in the Hilbert and Chapman-Enskog theories it is only the initial values which incorporate a lack of fluidity. There is an important qualitative distinction in that the shift in the level of description is in one case a rigorous consequence of a certain limit ($\varepsilon \to 0$) whereas it is in the other case arbitrarily imposed (but much more flexible).

Within classical fluid dynamics itself there are very many formulations ranging from compressible, viscous, and heat conducting equations governing a state defined by velocity and thermodynamic coordinates to potential flow, described by a single scalar function. The passage from one to another formulation, involving changes in the number of state variables and in the amount of initial and boundary data, are very singular and very complex. There are many rigorous examples showing how a sufficiently singular limit can change not only the level of description but also profound qualitative properties of solutions. Of course, the complexities are even greater in the transition to the microscopic Boltzmann and Liouville equations; but there is no reason to consider them more mysterious.

3. Redundant Hypotheses

For purposes of illustration we introduce this section with an extremely elementary example. There are probably good pedagogical reasons for introducing at an elementary level the "derivation" of the perfect gas law which proceeds from the assumption that one-third of the molecules are moving in the direction of each coordinate axis. One hopes that the pedagogue will not neglect to remark that in a more advanced course it will be shown that the same result follows without this drastic assumption.

A slightly more sophisticated bit of sleight-of-hand is the derivation of the Maxwellian velocity distribution by assuming that the molecular distribution is isotropic with statistically independent distributions in the three rectangular velocity components. The adoption of a most probable state as the *definition* of equilibrium in many texts is similarly divorced from dynamics.

The unfortunate fact is that very many arguments in statistical mechanics are of this type. They are obtained by magical devices using combinations of completely arbitrary and sometimes demonstrably false rules which have no connection with the established laws of physics. Their only salvation is that most applications are made in cases where we already know or believe we know the correct answer.

Turning to a more sophisticated example, equilibrium statistical mechanics is frequently founded on the postulate that equal *a priori* probability is assigned to equal volumes in phase space[1]. This is either an important natural law which supplements the equations of motion (but only in equilibrium), or it is redundant and can be dispensed with. We shall return to this question in Sec. 5 but consider here the special case of an ideal gas. The appropriate application is to the microcanonical distribution which places a uniform density on the surface of an energy sphere, $\frac{1}{2}m(v_1^2+v_2^2+\cdots v_N^2)=NE$. A simple geometrical calculation shows that the projected area of this sphere on one coordinate axis is approximately Maxwellian for large N. But it is not difficult to generalize this result to show that almost any distribution, replacing the uniform one on the energy sphere, will yield exactly the same result. More precisely, any distribution which is bounded by a constant independent of N, or even one whose maximum value increases more slowly than exponentially in N will suffice. In order to produce a non-Maxwellian distribution (e.g., as an initial value for a non-equilibrium solution of the Boltzmann equation), the probability density f_N must be concentrated on a fraction of the energy sphere which is exponentially small in N. Even to produce a true equilibrium distribution which is Maxwellian but has a finite angular momentum requires choosing an "extremely unlikely" distribution. This shows that a certain amount (how much?) of the dynamics must be taken into account before abandoning dynamics for probability.

We see now what the significance of the microcanonical assumption is. All but "extremely unlikely" distributions yield a Maxwellian for f_1. The microcanonical is just a special choice, more amenable than most to calculation. One lesson to be learned is that every nonequilibrium state is extremely unlikely. We must be extremely careful in dismissing possibilities which are unlikely without verifying, at least, that they are even more unlikely than the phenomena of interest. For example, many

---

[1] For example see Tolman [*14*].

arguments which presume to derive the Boltzmann equation turn out under more careful examination to be justified only for well-behaved distributions which are not wild enough to allow any state but the Maxwellian.

One example in which it is possible to distinguish between two distinct levels of unliklihood is the problem of molecular chaos. A simple version of this condition is

$$f_2(P_1, P_2) = f_1(P_1) f_1(P_2). \tag{3.1}$$

The two molecules are statistically independent. Some form of the molecular chaos hypothesis is employed in every derivation of the Boltzmann equation. On a purely probabilistic basis it can be shown that within the (unlikely) class of functions $f_N(P)$ which is compatible with a given non-Maxwellian $f_1(P_1)$, a very small part (more unlikely) fails to satisfy the chaos condition (3.1) when N is large [*15*]. This would seem to justify the chaos condition which is required for the Boltzmann equation. But direct application of (3.1) gives, not the Boltzmann equation, but identically zero for the collision term! What is required for the Boltzmann equation is chaos for particles which are approaching one another and a very definite finite correlation in that part of phase space corresponding to particles which have just collided. Examination of LIOUVILLE's equation for an initial f_N satisfying (3.1) shows that chaos is instantly destroyed over part of phase space. The equations of motion convert a likely state into one which is less likely; but this is just what is needed to justify the Boltzmann equation. We emphasize that the hypothesis that is needed to obtain the Boltzmann equation is not what is suggested by an estimate of the most likely state; it is this coupled with an important modification which is imposed by dynamics. The correct answer is intuitive in this case; the proper balance between dynamics and probability may not be so clear in other cases.

With regard to the casual dismissal of sets of small measure as being unlikely, it is interesting to note that the entire evolution of $f_1(P_1)$ over a macroscopic period of time is governed by values of $f_2(P_1, P_2)$ in a very small part of the two particle phase space. One of the important difficulties in presenting a careful derivation of BOLTZMANN's equation from LIOUVILLE's equation is to verify that the small set on which chaos is destroyed during the motion of the system does not interfere with another small set on which chaos is needed [*1*].

The concept of a functional dependence of a higher order description on lower order variables is a common expedient in reducing the level of description. This is a basic part of the Chapman-Enskog theory, and in this case the hypothesis can be justified as a rigorous asymptotic approximation in the limit of small mean free path [*12*]. This concept is also used by BOGOLUBOV to eliminate f_N in favor of f_1 in a derivation of BOLTZMANN and higher order kinetic equations [*10*]. The initial limit introduced by BOGOLUBOV ($N \rightarrow \infty$ keeping $N\sigma^3$ fixed in our notation) is not singular enough to produce such a drastic reduction in level. When this is followed by a power series in $N\sigma^3$, it becomes plausible that the functional hypothesis is valid for the first term or two, but simple estimates show that it cannot be valid to arbitrary order[1] (as it is in the Chapman-Enskog case).

In almost all treatments of statistical thermodynamics, the entire argument is based on one constant of the motion, the energy. We have already noted that, in a system which allows angular momentum, the correct state of the system is confined to a set of very small measure on an energy surface. We must show that there are no constants other than the energy in a given case, or correctly take into account the other integrals, or else show that they are somehow irrelevant. This will be discussed at length in Sec. 5.

Most formulations of statistical thermodynamics are also content to demonstrate that there is a formal similarity between the structure which is obtained and macroscopic thermodynamics when appropriate variables are identified. But the structure which arises in some statistical treatments can be shown to be non-unique according to the usual recipes. It is true that the usual identification is the correct one (otherwise it could not have survived comparison with experiment). But this should be shown to be a logical consequence of the laws of motion (cf. Sec. 5).

The subject of information theory has been used to give a very powerful formulation of nonequilibrium steady flows [*16*]. The essential point is to choose a distribution function which maximizes a certain well-chosen entropy. But this extraneous concept has not yet been connected in any way with dynamics. It can only be looked upon as an *ad hoc* recipe whose accuracy must be empirically determined. Either this hypothesis is correct, in which case it is unnecessary, or it is incorrect and should not be used. Just as in the case of chaos, we do not know in what small or crucial way the laws of dynamics may refuse to follow the most probable path. But, to prevent misunderstanding, we repeat that our objection is a matter of logical necessity; this method can be exceedingly valuable in practice while we wait for mathematics to catch up.

Summarizing, we see that there are frequent cases where the fact of large N dominates the scene, making the dynamical system look random and relegating the dynamics to a secondary role or at least simplifying the description. But this is not at all a universal phenomenon common to

---

[1] See footnote 1, p. 51.

all large systems. We must not be surprised if there are occasional errors created by an overeager use of probability. And even when the statistical hypothesis is a very good approximation to the actual state of affairs, it will usually be logically inconsistent with the exact dynamical formulation. It requires judicious care to separate valid consequences from paradoxes.

4. Reversibility

The relation between microscopic reversibility and macroscopic irreversibility has undoubtedly created more confusion than any other question in mathematical physics. The logical resolution of this question is quite elementary (although the mathematical details may be far from trivial). It is simply that one may approximate *arbitrarily closely* to an irreversible system by a reversible one and vice versa. A distinction between reversible and irreversible therefore has no physical content since physics cannot distinguish arbitrarily small differences. Whether a system is reversible or not depends on the choice of a mathematical model. Of course one may be much more convenient than another; but the distinction is mathematical and can be given no physical meaning. In particular, reversibility is not the antithesis of irreversibility. One can even find aspects of both in a single mathematical system. Unfortunately, although the logic of the question is easily resolved, the long history of sophistry makes the primary practical question a psychological one.

Consider first a simple collisionless gas in a rectangular box. The molecules collide with the walls but not with one another. The problem is dynamically reversible. Yet it is easily proved (with complete rigor) that for any smooth initial molecular distribution function, the density in physical space $\varrho(x, t)$ approaches a constant as $t \to \infty$. The same is true if we replace the initial distribution function by one with all velocities reversed. There is clearly no better indication of irreversibility than an approach to equilibrium. This problem is too simple to allow an approach to equilibrium in velocity as well as in physical space; the velocity of each particle is unchanged during the entire motion (except for reversals of sign). It is conjectured (but not proved) that for most domains more general than a rectangular box, the distribution will become isotropic in velocity as well as uniform in space; but the speed of each particle is still preserved.

The reason for the approach to a uniform state in this example is that a small change in initial velocity will produce a large change in position at a very much later time. Mathematically there is continuous dependence of the state of a molecule at time t on its initial values, but the measure of the degree of continuity becomes worse linearly in t. We state that the universal mechanism for irreversibility or superficially random behavior in a dynamical system is the progressive weakening with time of continuous dependence on initial conditions. Molecular collisions are a very effective mechanism because of the numerical values involved (cf. the numerical calculation described in Sec. 1). In various applications in physics this mechanism is found as phase mixing, Landau damping, random phase approximation, collisionless shocks, etc. But the universal mechanism is loss of continuous dependence.

Tossing a coin, spinning a roulette wheel, etc. all depend on the long times involved to make the outcome unpredictable. The necessity of mid-course corrections in sending a rocket to the moon is another aspect of the same effect. Whether a given situation is predictable or random may depend on the state of the experimental art. This is a well-known effect with dice, and it is also observed in the Spin-Echo experiment.

Perhaps the simplest mathematical model is given by the relation between a trigonometric series and a Fourier transform. Any convergent trigonometric sum, $\Sigma a_n \exp(i \omega_n t)$, finite or infinite, represents an almost periodic function. It keeps repeating essentially the same values and does not settle down as $t \to \infty$. This is true even if the numbers ω_n are dense in some interval. If the numbers ω_n are very close to one another, one might expect to approximate this sum by an integral, $\int a(\omega) \exp(i\omega t)\, d\omega$. For any reasonable function $a(\omega)$, this integral decays (irreversibly!) to zero as $t \to \infty$. It is true that the trigonometric sum can be made to approximate the integral arbitrarily closely in any finite time interval $0 < t < T$. But there is a nonuniform approach in the two limits $T \to \infty$ and taking many discrete approximating frequencies ω_n (i.e. many degrees of freedom). Under appropriate circumstances one can approximate arbitrarily closely to a discrete sum by a continuum and to a continuum by a discrete sum. But every discrete sum is almost periodic and every integral converges to zero as $t \to \infty$.

We now return to the problem of the gas of N particles in a box, this time allowing them to collide with one another. If, for simplicity, we take elastic spheres, the orbit of the system in $6N$-space is a sequence of straight lines, reflected specularly at a very complicated rigid boundary defined by the walls of the box and the various constraints $|x_i - x_j| = \sigma$. Qualitatively the problem can be expected to be the same as for specular reflection in the simple rectangular box. We can expect $f_1(v, x, t)$ [the projection of f_N into a subspace] to approach a limit as $t \to \infty$ just as $\varrho(x, t)$ [a projection of $f_1(v, x, t)$] approaches a limit in the rectangular box. This approach of f_1 to its equilibrium value does not depend on N being large; whether or not the limiting function is Maxwellian does depend on the size of N.

The Boltzmann equation predicts a very special type of approach to equilibrium; it is monotone when measured correctly. Neither the approach of $\varrho(x, t)$ to its limit nor the analogous approach of $f_1(v, x, t)$ to its limit when intermolecular collisions are included can be expected to be monotone. But we recall that the Boltzmann equation is found only after taking a certain limit in which $N\rightarrow\infty$ and $\sigma\rightarrow 0$. The continuous dependence estimate becomes infinitely poor in this limit, and the time required to completely forget the initial molecular state approaches zero. There are two evolutionary time scales, one of which is reached almost instantaneously as we approach the limit. It is this very singular limit which instantaneously reduces the relevant description from f_N to f_1 (this takes a finite time for finite N and σ) and at the same time makes the irreversible mechanism instantaneous and consequently monotone.

Conversely, it is exactly the extremely singular limit which makes rigorous estimates so hard. All the elaborate classical theory of ordinary differential equations is lost on a macroscopic time scale because of the loss of continuous dependence. And at the same time, as we have already mentioned, the evolution of the system is governed by its state on a set of vanishingly small relative measure.

We can with profit compare the transition from LIOUVILLE'S equation to BOLTZMANN'S equation in a singular limit which destroys continuous dependence in f_N and transfers the level of description to f_1 with the transition from BOLTZMANN'S equation to fluid equations in another singular limit which loses continuous dependence on f_1 (in a time ε, f_1 becomes locally Maxwellian) and transfers the level of description to the macroscopic state (in the latter example with complete mathematical proof).

It should be emphasized that the irreversibility that is inherent in the finite N-particle system (such as the approach to equilibrium) requires no recurrent stochastic manipulation of the system. If we were to numerically compute individual particle orbits in a system which is too large to maintain enough accuracy for "exact" orbits, the numerical truncation would amount to a small stochastic perturbation. It is well known that this type of numerical inaccuracy can be equivalent to an effective dissipation. Depending on the parameters, the numerical dissipation could be larger than the natural dissipation inherent in the exact system (in which case it is nonphysical), or it could be smaller than the natural dissipation, and one could consider that the macroscopic state has been computed exactly even if the orbits are not. These relative effects have not been estimated. The essential point is that although we cannot hope to locate individual molecules accurately enough to predict their individual motions, macroscopic laws (in particular irreversible behaviour) need not be based on this inaccuracy. Nothing need be altered in an N-particle dynamical system to obtain irreversibility; but to exhibit the irreversibility in a simple form without subtlety, it may be convenient to perform some stochastic manipulation, or preferably, to take an appropriate limit, $N \to \infty$.

To put to rest the psychological difficulties connected with irreversibility is impossible without some comments regarding entropy. The question is raised, how is it possible for entropy to be constant using LIOUVILLE's equation while entropy increases monotonely using BOLTZMANN's equation? An entirely analogous question is, how it is possible for the function $\sin x$ to be bounded while the function e^x is not? The answer is that the two entities which are given the same name "entropy" are quite distinct functions. There are in fact, very many different quantities which can be legitimately called entropy; for example $S_N = -\int f_N \log f_N \, dP$, $S_1 = -\int f_1 \log f_1 \, dP_1$, and many versions of thermodynamic entropy, S_T. They are all closely related (although not identical) in equilibrium, but they diverge otherwise.

We have seen that a single mathematical model can have both reversible and irreversible attributes; e.g. the orbits can be reversible while the density approaches equilibrium. Different entropies can be defined in a given model, some more relevant to the reversible coordinates and others descriptive of the irreversible features. For example, in the gas of molecules which collide with the walls alone, both S_N and S_1 are constant whereas S_T is not. In a real gas, obeying LIOUVILLE's equation, S_N is constant but S_1 and S_T are not. The latter do not vary monotonely, but each approaches a limiting value as $t \to \infty$, and under certain circumstances these limiting values can be shown to be larger than the initial values. For the Boltzmann equation, S_1 is monotone and S_T is not (S_N is not defined); but in the limit $\varepsilon \to 0$ in which fluid dynamics emerges, S_1 and S_T coalesce (except in boundary layers).

Roughly speaking, the highest order entropy (e.g. S_N) is relevant to a maximal reversible description. If all the particles are described precisely, then we always have complete information. A lower level, incomplete description may involve irreversible coordinates; this entropy is not constant, and is, in some circumstances, increasing. We must choose whichever entropy is appropriate to the variables and phenomena of interest, the degree of precision, and the length of time the system is observed.

Even in equilibrium there are several choices for S_T. Consider the following distinct experimental situations:

(1) two systems which can interchange thermal energy across an otherwise impermeable barrier;

(2) two systems which can interchange energy and momentum across a barrier which is free to move under molecular impact;

(3) two systems which are free to exchange energy and angular momentum;

(4) two systems connected by a capillary which allows the exchange of matter.

In all cases the total energy, momentum, angular momentum, and mass remain constant. The fact that these entirely different experimental situations turn out to be describable by *approximately* the same formal thermodynamic structure can be considered to be at least a minor miracle. Unfortunately this approximate agreement is usually distorted into exact conformity. Careful analysis, either on the macroscopic level or from statistical thermodynamics, shows that the thermodynamic structures are similar except for the entropy functions. This is an error (not a paradox) of GIBBS. The distinction between an assemblage of particles which are individually identified or indistinguishable can be made just as well in a classical as in a quantum analysis. If the particles are assigned labels (as they frequently are in a classical analysis), then the entropy appropriate to distinguished particles arises. But this procedure need not be followed. In experiment (4) above, the description of the experiment itself does not allow one to identify the molecules in one of the containers, and the undistinguished entropy automatically emerges when the calculation is done correctly. This is essentially the same function as the quantum entropy. If the correct entropy is used (and which one is correct is prescribed by the experiment and the nature of the observation), then it always increases when something substantive is changed. In particular, removal or insertion of a barrier in an otherwise homogeneous medium is substantive or not depending on whether the particles are distinguishable (which depends on the ability and interests of the experimenter).

In another context this concept of different thermodynamic entropies is very common. If a chemical reaction is slow, we have a choice of two distinct entropy functions. The more constrained description, in which the reaction is forbidden, is usually considered to be an approximation to the "exact" description. But this is no ordinary mathematical approximation in which a function is replaced by another whose values are close; the two entropy functions are completely different and are even functions of different variables.

In conclusion, we repeat that the fact that particle dynamics is in a certain sense reversible and macroscopic descriptions are sometimes irreversible is no barrier to their mutual correctness and compatibility (to a degree of approximation). Each level of description can exhibit simultaneous aspects of both reversibility and irreversibility, each aspect as well as each level with its own proper entropy function, some increasing, some being constant, and some exhibiting more complicated behavior[1].

[1] For a more complete account of the topics in this section see [*17*].

5. Equilibrium and the Approach to Equilibrium

No satisfactory theory of equilibrium can be formulated without basing it on an approach to equilibrium. Since the approach to equilibrium is a very difficult problem, it is frequently bypassed by plausible hypotheses which then allow the study of equilibrium *per se*.

To make the problem definite, let us consider an isolated system in a box, say with reflecting walls. As a representative initial value problem we can imagine that most of the gas is initially confined to half the box, and it is then released. It will take some macroscopic time τ_0 (seconds or minutes) to settle down to an empirical equilibrium state. This time scale is inherent in a dissipative fluid description, it emerges from the Boltzmann equation after suitable asymptotic analysis, and it is deeply hidden (but present) in an N-particle or Liouville description.

There are many possibilities for a definition of equilibrium. The simplest is to require that the state variables in a given description be unvarying in time. This is only sometimes useful. In a particle description where an observable is given by a phase function $\varphi(P)$, this definition cannot be used. By a theorem of POINCARÉ, the state variable P will ultimately come close to any value it has previously taken. The same is true for $\varphi(P)$. Either φ is an absolute constant (in the example above, even while the gas surges back and forth across the box), or it will never settle down. This is not just a mathematical difficulty; it is not clear that one can talk of equilibrium or an approach to equilibrium if there are, say, five particles in the box.

On the other hand, taking $N=10^{20}$ and $\varphi(P)$ as the fraction of molecules in the left half of the box, we fully expect φ to approach a limit. The mathematical theorem refutes this. Although it might take an astronomical time for φ to depart from its macroscopically expected value after this value has been approximately reached, it must eventually do so. In short, we must find a more practical definition of equilibrium to avoid events which are exceptional or rare in some sense.

On physical grounds we are led to expect an approach to equilibrium only for a macroscopic variable in a large system. The simplest way conceptually (but the hardest mathematically) to achieve this end is to let $N\to\infty$ before phrasing the question. This is exactly what is done in deriving the Boltzmann equation or a set of macroscopic fluid equations The approach to equilibrium emerges relatively simply, but only after taking a very sophisticated limit, $N=\infty$. To avoid the introduction of such a difficult and singular limit even before we define equilibrium, there are available two standard expedients. One is to introduce the concept of repeated experiments with an *a priori* probability $f_N(P)$ and the other is to consider infinite time averages. After considerable analysis the two paths are found to merge. But we shall see that this is not a trivial conclusion.

First consider a nonsingular solution $f_N(P, t)$ of LIOUVILLE's equation. Equilibrium is defined as a time-independent solution $f_N(P)$. What is stationary is not the value of an observable $\varphi(P)$ but its probability distribution. For example, let φ be the x-coordinate of a particular molecule. The expected value

$$\langle\varphi\rangle = \int \varphi(P) f_N(P)\, dP \tag{5.1}$$

will be at the center of the box. The distribution of values taken by φ will presumably be uniform over the box. But $\varphi(P)$ itself is a violently oscillating quantity. If $\varphi(P)$ is the number of molecules in the left half of the box, $\langle\varphi\rangle = \frac{1}{2}N$. The distribution of values about the mean will be proportionally small if N is large. In this case we could say that *the* observed value of φ/N is $\frac{1}{2}$.

By introducing repeated experiments and probability, we have succeeded in separating the study of equilibrium from the approach to equilibrium. If there is a true approach to equilibrium, we can compute the properties of the equilibrium. But we can compute even where no equilibrium exists in the accepted sense. For example, a macroscopically turbulent state which does not decay (on some time scale) could be compatible with a stationary probability $f_N(P)$. The fact that there actually is an approach to equilibrium in terms of f_N can be made plausible, but this conclusion is mathematically incomplete at the present time. We can expect *weak convergence*

$$f_N(P, t) \underset{\omega}{\longrightarrow} f_N^0(P) \tag{5.2}$$

in the sense that every expectation converges in the usual sense,

$$\langle\varphi\rangle = \int f_N(P, t)\, \varphi(P)\, dP \to \varphi^0 = \int f_N^0(P)\, \varphi(P)\, dP \tag{5.3}$$

and even the lower order distributions $f_1(P_1)$, $f_2(P_1, P_2)$, etc. converge pointwise to definite limits. The important point to remember is that convergence of the expected value of φ to a limit in no way contradicts the Poincaré recurrences of φ itself. As we have already mentioned, a rigorous proof of the approach to equilibrium can be given in the special case of non-colliding particles in a rectangular box (cf. Sec. 4). In this case $P(t)$ does not converge and exhibits Poincaré recurrences, but $f_1(P, t)$ converges weakly, and all observables $\langle\varphi\rangle$ as well as the lower order distribution $\varrho(x, t)$ (and the thermodynamic entropy) converge pointwise to constant values. There is no reason to doubt the weak convergence of f_N and pointwise convergence of f_1 for solutions of LIOUVILLE's equation in general, but the proof is an open question.

Taking the approach for granted, we return to the question of time-independent solutions $f_N(P)$ of LIOUVILLE's equation. Such a solution has the property that it is invariant under the transformation $P(0) \to P(t)$; in other words $f_N(P)$ is constant on every orbit $P(t)$. This is exactly the definition of a *time-independent integral* of the equations of motion (with the minor distinction that f_N is non-negative and normalized, $\int f_N dP = 1$). The question now is to find all time-independent integrals of the equations of motion.

In answer to this question, we find two extremes in the literature. A frequent statement which is made in texts on mechanics as well as statistical mechanics is that there exist $6N-1$ such integrals. This is trivially false. On the other hand, almost all classical statistical mechanics calculations are made on the basis of a single integral, the energy. This requires justification. It is a simple matter to show that there is no possible justification for statistical thermodynamics as a *prediction* from particle dynamics (as distinguished from an arbitrary formalism which empirically agrees with experiment) unless we know (or assume) that the energy is the only time-independent integral [*18, 19*].

There are many elementary examples to show that there need not exist a maximal set of $6N-1$ integrals. For a non-colliding gas in a rectangular box it is easily shown that any function $f(x, y, z, u, v, w)$ which is invariant under all trajectories $x(t)$, $y(t)$, $z(t)$, $u(t)$, $v(t)$, $w(t)$ must be a function of $|u|$, $|v|$, and $|w|$ alone. Thus there are three integrals instead of a maximum of five. Similarly, for a particle in almost any attractive central force (an exception is the inverse square law), there are only three time-independent integrals (energy and two components of angular momentum) instead of five.

One attempt to justify the use of energy alone in the possible presence of other integrals has been made by KHINCHIN [*20*] (and in a modified form by TRUESDELL [*21*]). The argument is that there is a large class of observables $\varphi(P)$ which are found to be almost constant over the entire energy surface except on a set of small relative measure (if N is large). But this argument is misleading. One cannot casually dismiss sets of small measure. In a system with finite macroscopic angular momentum, $f_N(P)$ is concentrated entirely on a set of exponentially small measure in N. The observables estimated by KHINCHIN differ from their "expected" values on exactly such small sets. One cannot be sure of predicting macroscopic results without verifying that no integrals have been disregarded. If there is an integral which is either ignored or unknown, the experimental results will depend on the accidental values given to these integrals.

Since any constant of the motion yields a quantitative distinction between two different macroscopic systems, it should be experimentally observable and therefore controllable. If it involves a phenomenon which does not couple into molecular motions (e.g., involving the nucleus), it can safely be ignored. If it is not entirely uncoupled from molecular motion, its neglect can be expected to produce an unexplained experimental scatter which should lead to its discovery.

The correct procedure is to generalize the usual approach which is based on the energy integral ε to one which takes into account a *complete set of integrals* $\varepsilon_1, \ldots, \varepsilon_r$ [*18, 19*]. Truly macroscopic conclusions follow only if r is a small number. This procedure is founded on the basic conjecture that the differential equations of dynamics have only a small number of integrals which are usually associated with evident symmetries. This conjecture has been strikingly confirmed by a proof by SINAI [*22*], that a system of N hard spheres in a rectangular box has no integrals other than the energy. There is no reason to doubt that this will also be true for more general force laws, but this is an important open question.

From a complete set of integrals $\varepsilon_1(P) \ldots \varepsilon_r(P)$, all equilibrium properties can be computed in terms of a function $g(\varepsilon_1, \ldots, \varepsilon_r)$ where

$$f_N(P) = g(\varepsilon_1 \ldots \varepsilon_r). \tag{5.4}$$

The results (i.e., expected values $\langle\varphi\rangle$) will depend on what is taken for the function g. Since the integrals are in principle controllable, an experiment can be done with the integrals taking arbitrarily assigned values $\varepsilon_i(P) = E_i$. More precisely, there will be some small spread about the desired value E_i, but continuity of the expected values with respect to E_i (this must be verified in a given case) makes this spread unimportant. We are therefore led in an isolated system, to the *generalized microcanonical distribution* on the manifold $\varepsilon_i = E_i$. If the experimental circumstances are such that there is a significant spread in the values of the ε_i, the calculation will depend on a knowledge of the exact distribution $g(\varepsilon_1 \ldots \varepsilon_r)$.

The second major approach is to introduce the infinite time average

$$\overline{\varphi} = \lim \frac{1}{T} \int_0^T \varphi\big(P(t)\big)\, dt \tag{5.5}$$

as the equilibrium value of the observable $\varphi(P)$. This procedure begs the question of the approach to equilibrium (as does the f_N formulation), and it does not even define a state of equilibrium but only "equilibrium values". But it does apply to a single isolated system and makes no immediate appeal to repeated experiments or probability.

This time average must not be confused with an average which is sometimes taken to smooth out microscopic fluctuations that would not be felt by a macroscopic instrument. For a general function $\varphi(P)$, the time required for the limit (5.5) to converge is astronomical, viz. the time it would take for P to sample the entire accessible phase space in $6N$-dimensions. If φ is symmetric in its arguments $P=(P_1 \ldots P_N)$, the time required for a reasonable estimate of $\overline{\varphi}$ could be expected to approximate the macroscopic equilibration time τ_0; this is the time required by a single molecule to sample its entire phase space P_1. Only for a spatially homogeneous system would the time T be microscopic, on the order of a number of collision times.

If $\varphi(P)$ were to approach a limit as $t \to \infty$, this limit would evidently have the value $\overline{\varphi}$. Since φ does not approach a limit, we must look for another justification for time averaging. The simplest is to consider t itself as a uniformly distributed random variable, thereby inducing a probability distribution on the values taken by $P(t)$ and $\varphi(P)$. Physically one can imagine that φ is observed at randomly selected times. This is an artificial introduction of probability since the entire trajectory $P(t)$ is in principle known after one complete observation. This is to be distinguished from the previous introduction of f_N which postulated a lack of information about the initial values $P(0)$.

By the decision to interpret t as a random variable, the formula (5.5) for $\overline{\varphi}$ defines the expected value of φ. By considering the random variable $\varphi_a(P)$ defined as

$$\left.\begin{aligned} \varphi_a(P) &= 1 \quad \text{if} \quad \varphi(P) > a, \\ \varphi_a(P) &= 0 \quad \text{if} \quad \varphi(P) \leqq a, \end{aligned}\right\} \tag{5.6}$$

we can evaluate the entire probability distribution of the random variable φ as $\overline{\varphi}_a$. We repeat that the actual time interval T required to obtain an expected value is not physically relevant. If T is one second, the average $(1/T)\int \varphi\, dt$ will vary significantly from second to second. If T is one hour, the average will fluctuate significantly if we wait a very long time. And as we shall see, if the initial point P is chosen poorly, the expected values of any observable will have no relation to any experimental observation.

To see exactly what the time average implies we refer to the ergodic theorem which states that the time average $\overline{\varphi}$ exists (almost everywhere) and is equal to a phase average over a properly chosen domain. To make this a little more precise, we first define an invariant set as one which contains an entire trajectory $P(t)$ if it contains the initial point, and define an indecomposible set as an invariant set which cannot be subdivided into two invariant subsets each of positive measure; (for almost every initial point in an indecomposible set the corresponding trajectory will be dense in the set). The time average will equal a phase average on an appropriate indecomposible set.

We have obtained a complete but at the same time enormously complicated answer to the question of evaluating time averages. An energy shell $E' < \varepsilon < E''$ is a $6N$-dimensional invariant set. An energy surface $\varepsilon = E$ is a $(6N-1)$-dimensional invariant set. Similarly the intersection of any number of integral surfaces $\varepsilon_1 = E_1, \ldots, \varepsilon_r = E_r$, is a $(6N-r)$-dimensional invariant set. Any periodic motion is a 1-dimensional invariant set. In a rectangular box we can set each molecule in motion perpendicular to a wall in such a way that molecules do not collide with one another; if the transit times are incommensurable, we have an N-dimensional invariant set. There is a veritable jungle of invariant sets and therefore of possibilities for the evaluation of the time average $\overline{\varphi}$.

The difficulty can be formulated in another way. Whether a set is indecomposible or not depends not only on the set but on a decision as to the nature of the underlying measure. Implicit in the above definitions is that we are given a measure which is invariant under the transformation $P(0) \to P(t)$. But this is far from unique. One such invariant measure is the $6N$-dimensional Lebesgue measure (LIOUVILLE's theorem). This measure can be used to induce a singular invariant measure on an energy surface or on any intersection of integral surfaces $\varepsilon_i = E_i$ (generalized microcanonical measure). With the full $6N$-measure, there are no indecomposible sets (an energy surface has measure zero). With a microcanonical measure, the entire integral surface is indecomposible if the set of integrals is complete. Any of the lower dimensional invariant sets can be assigned an invariant measure. And any indecomposible invariant set is a possible choice on which to evaluate a phase average.

To cut our way through the jungle of invariant sets and invariant measures we appeal to the principle that an observation must be reproducible. It must be possible to repeat an experiment in another laboratory and confirm substantially the same result. We can call this a law of physicists rather than a law of physics. Applying this criterion eliminates any *isolated* low dimensional invariant sets. More precisely, any invariant set of dimension less than $6N$ is eliminated if it is not the limit of a contracting sequence of $6N$-dimensional invariant sets. The time average over a periodic orbit contained within a certain energy surface is different from the time average belonging to a neighboring initial point which is ergodic on the entire surface. But a microcanonical average is changed only slightly if the initial point is taken on a neighboring energy surface.

The requirement of continuity of $\overline{\varphi}$ with respect to the phase point P is not predicated on the introduction of probability, even though both are motivated by the impossibility of exactly reproducing an initial state. If an arbitrarily small change in the position of a single molecule changes the value of $\overline{\varphi}$ by a finite amount, then this value of $\overline{\varphi}$ is not observable. Only if continuity were violated for every phase point P would we be forced to introduce probability. But even if we do take refuge in probability to eliminate the tangle of lower-dimensional invariant sets, we make only very small use of it. The applicable statement is that the set of points P which do not lie on trajectories which are ergodic on some microcanonical manifold is of Lebesgue measure (i.e., probability) zero. We recall that the concept of measure zero is much more primitive than that of finite measure.

This requirement of reproducibility brings the definition of a time-averaged observable into agreement with the microcanonical distribution as obtained from stationary f_N for a complete set of integrals. The definition of equilibrium in terms of f_N leads to the microcanonical result more quickly than through time averages and without any reference at all to ergodicity. The greater complication of the time average approach is inherent in its concentration on a single system rather than a probability density. But even though the latter does refer to a single system, it does not give direct information about φ itself but only about time averages and correlations. The equivalence of the two approaches shows that the sample space introduced by successive observations of a single isolated system is essentially the same as the sample space which is obtained by repeating an experiment while holding a complete set of integrals fixed. [This is a weaker statement than a hypothesis which is sometimes made: that the fluctuations (on a relatively short time scale) of a single system are similar to the dispersion induced by a distribution f_N of initial values.]

The concept of a complete set of integrals and generalized microcanonical distribution, introduced by GRAD [*18*, *19*], has been made precise in an elegant analysis by R. M. LEWIS [*23*] employing the framework of ergodic theory. The latter analysis takes the time average as a starting point and is formulated in terms of an unspecified given invariant measure and an associated complete set of integrals. But there is no unique choice of an invariant measure for a given dynamical system, and there is a different set of integrals with each measure. For example, consider a case in which energy is the only integral (with respect to Lebesgue measure). We can find a one-parameter family of periodic solutions, one for each energy surface. These have Lebesgue measure zero. But a legitimate invariant measure is obtained by assigning a total measure $\frac{1}{2}$ to a $6N$-dimensional energy shell and the remaining measure $\frac{1}{2}$

to the one-parameter family of periodic curves. With regard to this measure there are two distinct integrals in a complete set. In order to *predict* physical behavior rather than merely construct a formulation which has the same structure as macroscopic thermodynamics, we must appeal to the concept of reproducibility of an observation which restricts the allowable measures to the unique $6N$-dimensional Lebesgue measure and its microcanonical constructs.

To derive thermodynamics (or more properly thermostatics) it is necessary to consider systems which are weakly coupled. The basic conjecture is that even an arbitrarily small coupling between two systems is enough to destroy the individual integrals and preserve only the integrals of the combined system. On the other hand, microcanonical averages on a surface $\varepsilon = \text{const.}$ in the combined phase space are continuous in the coupling parameter. In other words, the equilibrium properties of weakly coupled systems can be evaluated by using the isolated Hamiltonians to compute a total energy surface. Of course the time required to reach equilibrium will depend sensitively on the coupling parameter even though the resulting equilibrium will not.

The existence of a unique temperature (one for each integral ε_i), the fact that any dynamical system can be used as a thermometer to measure the temperature of another large enough system, the existence of an entropy function with the needed convexity in its arguments, and the validity of the second law of thermodynamics all follow as derived consequences, not just by arbitrary identification of interesting parameters in the microscopic theory with macroscopic counterparts [*18*, *19*, *17*]. There are a variety of physical situations that can be treated depending on which of the integrals ε_i are taken to be weakly coupled and which are kept isolated.

One can also relax constraints involving parameters in the Hamiltonian such as the position of a separating wall; and one can relax the fixed number of molecules, either by removing a wall, or by allowing a previously forbidden chemical reaction to take place. This leads to a large variety and combination of canonical and grand canonical distributions. These different distributions arise naturally as descriptions of different physical situations and exhibit different thermodynamic structures. There is a valid point in looking for an abstract thermodynamic structure which is general enough to comprehend a large variety of physical situations. But one should not lose sight of the fact that a more precise examination reveals the necessity for a variety of structures.

The simple thermodynamic structures with unambiguous temperatures, etc., arise only in the limit of large N. A more subtle physical property in the limit of large N is that the volume is a thermodynamic parameter independent of the shape of the container [*24*]. But this question is only mathematically difficult and does not appear to have given rise to any psychological blocks.

There is a very interesting and elegant alternative derivation of the canonical and grand canonical distributions which shows directly how insensitive the result is to any details of molecular model, etc., by using statistical estimation theory [*25*]. The model is stochastic and proceeds from the macroscopic experimental observation that the statistic is "sufficient"; this empirical property is equivalent to the dynamical assertion that the energy is the only integral. This theory does give not an alternative method for deriving macroscopic laws from microscopic. But it does provide a very general extension of the classical macroscopic thermodynamic structure to include canonical fluctuations on a strictly phenomenological basis.

With regard to the approach to equilibrium only the case of the collisionless gas is mathematically complete; and only in the limit of a perfect gas (Boltzmann limit) can one make plausible estimates which have the flavor of mathematical proof. Even proof of the conjecture that, in general, f_N approaches a weak limit would not give a complete answer without a quantitative estimate of the time that is required. There will be several different relaxations on completely different time scales, and if some relaxation is astronomically slow (e.g., a very slow chemical reaction), then it is not the simple mathematical limit of f_N as $t \to \infty$ that is physically relevant. The more satisfactory alternative, proof of the approach to equilibrium in a single system after letting $N \to \infty$, is probably less accessible mathematically than one involving LIOUVILLE's theorem and f_N.

6. Conclusion

Classical physics is traditionally presented as a collection of fragments: thermodynamics, statistical thermodynamics, kinetic theory, particle dynamics, fluid dynamics, etc. The transition from one representation to another is dealt with more by art than by science, although in a few cases a sound mathematical foundation is available. In practice, the transitions are made by discarding a selected portion of the physical laws, replacing them by hypotheses which are logically extraneous. In the most interesting cases a shift in the level of description is mandatory because the alternative description becomes infinitely poor in some limit. The emergence of a deterministic lower order description is a truly surprising phenomenon without which there could be no macroscopic physics. It is not clear in some cases (turbulence, kinetic theory of dense media, "many body" problems) whether we have yet to find the correct low order description, or whether there may be none. For example, it could turn out that all but the gross features of turbulent flows depend on unmeasurable properties of the initial and boundary data. The belief in macroscopic causality is an act of faith more than of reason.

REFERENCES

[1] GRAD, H.: Principles of the kinetic theory of gases. In: Handbuch der Physik, vol. 12, p. 205. Berlin-Göttingen-Heidelberg: Springer 1958.
[2] WEINSTOCK, J.: Phys. Rev. **132**, 454 (1963).
[3] SANDRI, G.: Ann. Phys. (N.Y.) **24**, 332 (1963).
[4] DORFMAN, J. R., and E. G. D. COHEN: Phys. Letters **16**, 125 (1965).
[5] SENGERS, J. V.: Phys. Rev. Letters **15**, 515 (1965).
[6] ANDREWS, F. C.: Phys. Letters **21**, 170 (1966).
[7] GOLDMAN, R., and E. A. FRIEMAN: To appear in J. of Mathematical Physics.
[8] KAC, M.: Probability and related topics in physical sciences. New York: Interscience Publ. 1959.
[9] MAYER, J. E.: J. Chem. Phys. **33**, 484 (1960).
[10] BOGOLUBOV, N. N.: J. Phys. U.S.S.R. **10**, 265 (1946).
[11] MORREY, C. B.: Comm. Pure and Appl. Math. 8, 279 (1955).
[12] GRAD, H.: Phys. Fluids **6**, 147 (1963).
[13] — Comm. Pure and Appl. Math. **2**, 331 (1949).
[14] TOLMAN, R. C.: The principles of statistical mechanics. Oxford 1938.
[15] GRAD, H.: J. Chem. Phys. **33**, 1342 (1960).
[16] JAYNES, E. T.: Information theory and statistical mechanics. In: Statistical physics, pp. 181—218. New York: W. A. Benjamin Publ. 1963. [1962 Brandeis Summer Institute Lectures, vol. 3.]
[17] GRAD, H.: Comm. Pure and Appl. Math. **14**, 323 (1961).
[18] — Comm. Pure and Appl. Math. **5**, 455 (1952).
[19] — J. Phys. Chem. **56**, 1039 (1952).
[20] KHINCHIN, A. I.: Statistical mechanics. Dover 1949.
[21] TRUESDELL, C. A.: Six lectures on modern natural philosophy. Berlin-Heidelberg-New York: Springer 1966.
[22] SINAI, JA. G.: Doklady Akad. Nauk S.S.S.R. **153**, No 6, 1261 (1963).
[23] LEWIS, R. M.: Arch. Rational. Mech. Anal. **5**, 24 (1960).
[24] HOVE, L. VAN: Physica **15**, 951 (1949).
[25] MANDELBROT, B.: Compt. rend. **243**, 1835 (1956).

Chapter 6

Foundations of Probability Theory and Statistical Mechanics

EDWIN T. JAYNES

Department of Physics, Washington University
St. Louis, Missouri

1. What Makes Theories Grow?

Scientific theories are invented and cared for by people; and so have the properties of any other human institution — vigorous growth when all the factors are right; stagnation, decadence, and even retrograde progress when they are not. And the factors that determine which it will be are seldom the ones (such as the state of experimental or mathematical techniques) that one might at first expect. Among factors that have seemed, historically, to be more important are practical considerations, accidents of birth or personality of individual people; and above all, the general philosophical climate in which the scientist lives, which determines whether efforts in a certain direction will be approved or deprecated by the scientific community as a whole.

However much the "pure" scientist may deplore it, the fact remains that military or engineering applications of science have, over and over again, provided the impetus without which a field would have remained stagnant. We know, for example, that ARCHIMEDES' work in mechanics was at the forefront of efforts to defend Syracuse against the Romans; and that RUMFORD's experiments which led eventually to the first law of thermodynamics were performed in the course of boring cannon. The development of microwave theory and techniques during World War II, and the present high level of activity in plasma physics are more recent examples of this kind of interaction; and it is clear that the past decade of unprecedented advances in solid-state physics is not entirely unrelated to commercial applications, particularly in electronics.

Another factor, more important historically but probably not today, is simply a matter of chance. Often, the development of a field of knowledge has been dependent on neither matters of logic nor practical applications. The peculiar vision, or blindness, of individual persons can be decisive for the direction a field takes; and the views of one man can persist for centuries whether right or wrong. It seems incredible to us today that the views of Aristotle and Ptolemy could have dominated thought in mechanics and astronomy for a millenium, until GALILEO and others pointed out that we are all surrounded daily by factual evidence to the contrary; and equally incredible that, although thermometers (or rather, thermoscopes) were made by GALILEO before 1600, it required another 160 years before the distinction between temperature and heat was clearly recognized, by JOSEPH BLACK. (Even here, however, the practical applications were never out of sight; for GALILEO's thermoscopes were immediately used by his colleagues in the medical school at Padua for diagnosing fever; and JOSEPH BLACK's prize pupil was named JAMES WATT). In an age averse to any speculation, FRESNEL was nevertheless able, through pure speculation about elastic vibrations, to find the correct mathematical relations governing the propagation, reflection, and refraction of polarized light a half-century before MAXWELL's electromagnetic theory; while at the same time the blindness of a few others delayed recognition of the first law of thermodynamics for forty years.

Of far greater importance than these, however, is the general philosophical climate that determines the "official" views and standards of value of the scientific community, and the degree of pressure toward conformity with those views that the community exerts on those with a tendency to originality. The reality and effectiveness of this factor are no less great because, by its very nature, individual cases are more difficult to document; its effects "in the large" are easily seen as follows.

If you make a list of what you regard as the major advances in physical theory throughout the history of science, look up the date of each, and plot a histogram showing their distribution by decades, you will be struck immediately by the fact that advances in theory do not take place independently and randomly; they have a strong tendency to appear in small close clusters, spaced about sixty to seventy years apart. What we are observing here is the result of an interesting social phenomenon; this pressure toward conformity with certain officially proclaimed views, and away from free speculation, is subject to large periodic fluctuation. The last three cycles can be followed very easily, and the pressure maxima and minima can be dated rather precisely.

At the point of the cycle where the pressure is least, conditions are ideal for the creation of new theories. At these times, no one feels very sure just where the truth lies, and so free speculation is encouraged. New ideas of any kind are welcomed, and judged as all theories ought to be judged; on grounds of their logical consistency and agreement with experiment. Of course, we are only human; and so we also have a strong preference for theories which have a beautiful simplicity of concept. However, as stressed by many thinkers from OCCAM to EINSTEIN, this instinct seldom leads us away from the truth, and usually leads us toward it.

Eventually, one of these theories proves to be so much more successful than its competitors that, in a remarkably short time the pressure starts rising, all effective opposition ceases, and only one voice is heard. A well-known human frailty — overeagerness of the fresh convert — rides rough-shod over all lingering doubts, and the successful theory hardens into an unassailable official dogma, whose absolute, universal, and final validity is proclaimed independently of the factual evidence that led to it. We have then reached the peak of the pressure cycle; a High Priesthood arises whose members believe very sincerely that they are, at last, in possession of Absolute Truth, and this gives them the right and duty to combat errors of opinion with all the forces at their command. Exactly the same attitude was responsible, in still earlier times, for the Spanish Inquisition and the burning of witches.

At times of a pressure maximum, all free exercise of the imagination is frowned upon, and if one persists, severely punished. New ideas are judged, not on grounds of logic or fact, but on grounds of ideological conformity with the official dogma. To openly advocate ideas which do not conform is to be branded a crackpot and to place one's professional career in jeopardy; and very few have the courage to do this. Those who are students at such a time are taught only one view; and they miss out on the give and take, the argument and rational counter-argument, which is an essential ingredient in scientific progress. A tragic result is that many fine talents are wasted, through the misfortune of being born at the wrong time.

This high-pressure phase starts to break up when new facts are discovered, which clearly contradict the official dogma. As soon as one such fact is known, then we are no longer sure just what the range of validity of the official theory is; and we usually have enough clues by then so that additional disconcerting facts can be found without difficulty. The voice of the High Priests fades, and soon we have again reached a pressure minimum, in which nobody feels very sure where the truth lies and new suggestions are again given a fair hearing, so that creation of new theories is again socially possible.

Let us trace a few cycles of this pressure fluctuation (see Fig. 1). The pressure minimum that occurred at the end of the eighteenth century is now known as the "Age of Reason".

During a fairly short period many important advances in physical theory were made by such persons as LAPLACE, LAGRANGE, LAVOISIER, and FOURIER. Then a pressure maximum occurred in the first half of the nineteenth century, which is well described in some thermodynamics textbooks, particularly that of EPSTEIN [1]. This period of hostility toward free speculation seems to have been brought about, in part, by the collapse of SCHELLING'S *Naturphilosophie*, and its chief effect was to delay recognition of the first law of thermodynamics for several decades. As already noted, FRESNEL was one of the very few physicists who escaped this influence sufficiently to make important advances in theory.

Another pressure minimum was reached during the third quarter of the nineteenth century, when a new spurt of advances took place in a period of only fifteen years (1855—1870), in the hands of MAXWELL, KELVIN, HERTZ, HELMHOLTZ, CLAUSIUS, BOLTZMANN, and several

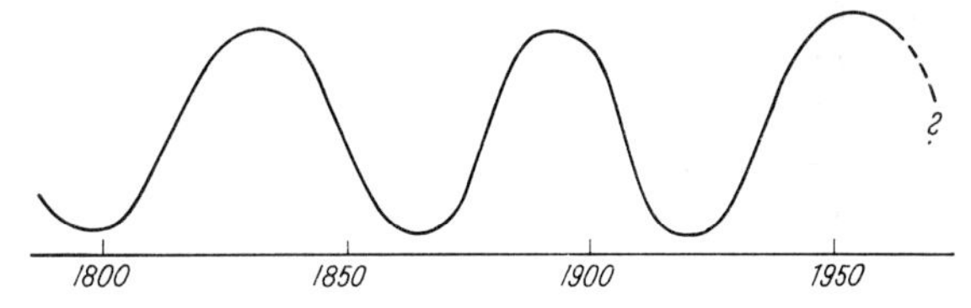

Fig. 1. Some recent fluctuations in social pressure in science

others. During this short period thermodynamics, electromagnetic theory, and kinetic theory were developed nearly to their present form; but the very success of these efforts led to another of the inevitable pressure maxima, which we recognize as being in full flower in the period 1885—1900. One of the tragedies (at least from the standpoint of physics) caused by this was the virtual loss of the talents of POINCARÉ. While his contributions to physical theory are considerable, still they are hardly commensurate with what we know of his enormous abilities. This was recognized and explained by E. T. BELL [2] in these words: "He had the misfortune to be in his prime just when physics had reached one of its recurrent periods of senility." The official dogma at that time was that all the facts of physics are to be explained in terms of Newtonian mechanics; particularly that of particles interacting through central forces. Herculean efforts were made to explain away MAXWELL'S electromagnetic theory by more and more complicated mechanical models of the ether — efforts which remind us very much of the earlier single-minded insistence that all the facts of astronomy must be explained by adding more and more Ptolemaic epicycles.

An interesting manifestation toward the end of this period was the rise of the school of "Energetics", championed by MACH and OSTWALD, which represents an early attempt of the positivist philosophy to limit the scope of science. This school held that, to use modern terminology, the atom was not an "observable", and that physical theories should not, therefore, make use of the concept. The demise of this school was brought about rapidly by PERRIN's quantitative measurements on the Brownian motion, which verified EINSTEIN's predictions and provided an experimental value for AVOGADRO's number.

The last "Golden Age of Theory" brought about by the ensuing pressure minimum, lasted from about 1910 to 1930, and produced our present general realitivity and quantum theories. Again, the spectacular success of the latter — literally thousands of quantitatively correct predictions which could not be matched by any competing theory — brought about the inevitable pressure rise, and for twenty-five years (1935—1960) theoretical physics was paralyzed by one of the most intense and prolonged high-pressure periods yet recorded. During this period the official dogma has been that all of physics is now to be explained by prescribing initial and final state vectors in a Hilbert space, and computing transition matrix elements between them. Any attempt to find a more detailed description than this stood in conflict with the official ideology, and was quickly suppressed without any attempt to exhibit a logical inconsistency or a conflict with experiment; this time, a few individual cases can be documented [*3*].

There are now many signs that the pressure has started down again; several of the supposedly universal principles of quantum theory have been confronted with new facts, or new investigations, which make us unsure of their exact range of validity. In particular, one of the fundamental tasks of any theory is to prescribe the class of physical states allowed by Nature. In MAXWELL's electromagnetic theory, for example, any mathematical solution of MAXWELL's equations is held to represent a possible physical state, which could in principle be produced in the laboratory. In quantum theory, we were taught for many years that the class of possible physical states is in 1:1 correspondence with solutions of the Schrödinger equation that are either symmetric or antisymmetric under permutations of identical particles. Our confidence in the universal validity of this rule has, recently, been shaken in two respects. In the first place, study of "parastatistics" has shown that much more general types of symmetry in configuration space can also be described by the machinery of quantized wavefunctions, and these new possibilities are not ruled out by experimental evidence. Secondly, the superposition principle (which may be regarded as a consequence of the above-mentioned rule, although it is usually considered in a still more general sense) holds that, if ψ_1 and ψ_2 are any two possible physical states, then any linear combination $\psi \equiv a_1 \psi_1 + a_2 \psi_2$ is also a possible physical state. But with the appearance of superselection rules, we are no longer sure what the range of validity of the superposition principle is.

The discovery of parity nonconservation was a great psychological shock; a principle which had been taught to a generation of physicists as a universally valid physical law, so firmly established that it could be used to rule out *a priori* certain theoretical possibilities, such as WEYL'S twocomponent relativistic wave equation, was found not to be universally valid after all; and again we are unsure as to its exact range of validity, and WEYL'S equation has been resurrected.

Several quantum mechanics textbooks assure us that the phenomenon of spontaneous emission places a fundamental irreducible minimum value on the width of spectral lines. Such statements are now confronted with the laser, which — in instruments now commercially available, and as simple to operate as a sixty-watt light bulb — produce spectral lines over a million times narrower than the supposedly fundamental limit! Thus, all around the edges of quantum theory we see the familiar kind of crumbling which, historically, has always signalled the incipient breakdown of the theory itself.

I hasten to add that, of course, none of these developments affects the basic "hard core" of quantum theory in any way; they show only that certain gratuitous additions to quantum theory (which had, however, become very closely associated with the basic theory) were unsound in the sense that they were not of *universal* validity. But it is inevitable that, faced these developments, more and more physicists will ask themselves how many other principles are destined to crumble a little at the edges, so that they can again be considered valid objects for inquiry; and not articles of faith to be asserted dogmatically for the purpose of discouraging inquiry.

In particular, the uncertainty principle has stood for a generation, barring the way to more detailed descriptions of nature; and yet, with the lesson of parity still fresh in our minds, how can anyone be quite so sure of its universal validity when we note that, to this day, it has never been subjected to even one direct experimental test?

Today, elementary particle theorists are busily questioning and reexamining all the foundations of quantum field theory, in a way that would have been regarded as utter heresy ten years ago; and some have suggested that perhaps the whole apparatus of fields and Hamiltonians ought to be simply abandoned in favor of more abstract approaches. It would be quite inconsistent with the present mood of theoretical physics if we failed to question and re-examine *all* of the supposedly sacred principles of quantum theory.

For all these reasons, I think we are going to see a rapid decrease in pressure in the immediate future, and another period of great theoretical advances will again be socially possible in perhaps ten years. And I think we can predict with confidence that some of the clues which will lead to the next round of advances are to be found in the many suggestions already made by dissenters from the Copenhagen theory — suggestions which have, thus far, been met only by sneers and attacks, which no attempt to study their real potentialities.

2. Statistical Mechanics

At this point, I see that you are looking about anxiously and wondering if you are in the right room; for the announced title of this talk was, "Foundations of Probability Theory and Statistical Mechanics". What has all this to do with statistical mechanics? Well, I wanted to say a few things first about general properties of physical theories because statistical mechanics is, in several respects, an exceptional case. Statistical methods exist independently of physical theories, and so statistical mechanics is subject to additional outside interactions from other fields. The field of probability and statistics is also subject to periodic fluctuations, but they are not in phase with the fluctuations taking place in physics (they are right now at a deep pressure minimum); and so the history of statistical mechanics is more complicated.

In particular, statistical mechanics missed out on the latest pressure minimum in physics, because it coincided with a pressure maximum in statistics; the transition to quantum statistics took place quietly and uneventfully without any real change in the basic formalism of GIBBS, and without any extension of the range of applicability of the theory. There was no advance in understanding, as witnessed by the fact that debates about irreversibility continue to this day, repeating exactly the same arguments and counter-arguments that were used in the time of BOLTZMANN; and the newest and oldest textbooks you can find hardly differ at all in their presentation of fundamentals. In short, statistical mechanics has suffered a period of stagnation and decadence that makes it unique in the recent history of science.

A new era of active work in statistical mechanics started, however, about 1955, in phase with a revolution in statistical thought but not at first directly influenced by it. This was caused, in part, by practical needs; an understanding of irreversible processes became increasingly necessary in chemical and mechanical engineering as one demanded more efficient industrial processing plants, stronger and more reliable materials, and bigger and better bombs. There is always a movement of scientific talent into areas where generous financial support is there for the taking. Another cause was the appearance of a few people who were genuinely interested in the field for its own sake; and perhaps it helped to reflect that, since it had been virtually abandoned for decades, one might be able to work in this field free of the kind of pressure noted above, which was paralyzing creative thought in other areas of physics.

Regardless of the reasons for this renewed activity, we have now made considerable progress in theoretical treatment of irreversible processes; at least in the sense of successful calculation of a number of particular cases. It is an opportune time to ask whether this has been accompanied by any better understanding, and whether the foundations of the subject can now be put into some kind of order, in contrast to the chaos that has persisted for almost a century. I hope to show now that the answer to both of these questions is yes; and that recent developments teach us an important lesson about scientific methodology in general.

Let me state the lesson first, and then illustrate it by examples from statistical mechanics. It is simply this: *You cannot base a general mathematical theory on imprecisely defined concepts. You can make some progress that way; but sooner or later the theory is bound to dissolve in ambiguities which prevent you from extending it further.* Failure to recognize this fact has another unfortunate consequence which is, in a practical sense, even more disastrous: *Unless the conceptual problems of a field have been clearly resolved, you cannot say which mathematical problems are the relevant ones worth working on; and your efforts are more than likely to be wasted.* I believe that, in this century, thousands of man-years of our finest mathematical talent have been lost through failure to understand this simple principle of methodology; and this remark applies with equal force to physics and to statistics.

2.1. BOLTZMANN's Collision Equation

Let us consider some case histories. BOLTZMANN sought to describe the approach to equilibrium in a gas in terms of the distribution $f(x, p, t)$. In his first work, this function was defined as giving the actual number of particles in various cells of phase space; thus if R denotes the set of points comprising a region of six-dimensional phase space, the number of particles in R is to be computed from

$$n_R = \int_R f(x, p, t)\, d^3x\, d^3p. \tag{1}$$

After some physical arguments which need not concern us here, BOLTZMANN concluded that the time evolution of the gas should be described by his famous "collision equation",

$$\frac{\partial f}{\partial t} + \sum_\alpha \left[\frac{P_\alpha}{m}\frac{\partial f}{\partial x_\alpha} + F_\alpha \frac{\partial f}{\partial P_\alpha}\right] = \int \partial^3 P' \int \partial \Omega\, (\bar{f}\bar{f}' - f f')\, \sigma \tag{2}$$

where F_α is the α-component of external force acting on a particle; and the right-hand side represents the effects of collisions in redistributing particles in phase space, in a way familiar to physicists. As a consequence of this equation, it is easily shown that the quantity

$$H_B \equiv \int f \log f \, d^3x \, d^3p \tag{3}$$

can only decrease (in this equation we integrate over all the accessible phase space); and so BOLTZMANN sought to identify the quantity

$$S_B \equiv -k H_B \tag{4}$$

with the entropy, making the second law of thermodynamics a consequence of the dynamical laws, as expressed by (2). As we know, this was challenged by ZERMELO and LOSCHMIDT who produced two counter-examples, based on time-reversal and on the POINCARÉ recurrence theorem, showing that Eq. (2) could not possibly be an exact expression of the dynamical equations of motion, *and thereby placing the range of validity of Boltzmann's theory in doubt.*

At this point, confusion entered the subject; and it has never left it. For BOLTZMANN then retreated from his original position, and said that he did not intend that $f(x, p, t)$ should represent necessarily the *exact* number of particles in various regions [indeed, it is clear that the only function f which has exactly the property of Eq. (1) is a sum of delta-functions: $f(x, p, t) = \Sigma_i \, \delta(x - x_i) \, \delta(p - p_i)$, where $x_i(t)$, $p_i(t)$ are the position and momentum of the i-th particle]. It represents only the *probable* number of particles; or perhaps the *average* number of particles; or perhaps it gives the *probability* that a given particle is to be found in various regions. The decrease in H_B is then not something which must happen *every* time; but only what will *most probably* happen; or perhaps what will happen *on the average*, etc.

Unfortunately, neither BOLTZMANN nor anybody else has ever become more explicit than this about just what BOLTZMANN'S f; and therefore BOLTZMANN'S H-theorem, means. When our concepts are not precisely defined, they are bound to end up meaning different things to different people, thus creating rooom for endless and fruitless debate, of exactly the type that has been going on ever since. Furthermore, when we debate about imprecise concepts, we can never be sure whether we are arguing about a question of fact; or only a question about the meaning of words. From BOLTZMANN'S day to this, the debate has never been able to rise above this level.

If you think my characterization of the situation has been too laconic, and unfair to many honest seekers after the truth, I invite you to examine a recent review article on transport theory [*4*]. On page 271, the author states that "The Boltzmann distribution function — is the (probable) number of particles in the positional range d^3x and the velocity range $d^3 v$ —". On page 274 this is altered to: "The quantity f, the Boltzmann distribution function — is, roughly speaking, the average number of particles in a cell in the $x-v$ space (the μ-space). f refers to a single system. A more precise definition of f can be obtained through the use of the master function P." Consulting this master function, we find that neither the definition of P, nor its connection with f, is ever given. This, furthermore, is not a particularly bad example; it is typical of what one finds in discussions of BOLTZMANN's theory.

Let us note some of the difficulties that face the practical physicist because of this state of utter confusion with regard to basic concepts. Suppose we try to assess the validity of BOLTZMANN's equation (2) for some particular problem; or we try to extend it to higher powers in the density, where higher order collisions will become important in addition to the binary ones that are taken into account, in some sense, in (2). If we agree that f represents an *average* number of particles, we must still specify what this average is to be taken over. Is it an average over the particles, an average over time for a single system, an average over many copies of the single system, or an average over some probability distribution? Different answers to this question are going to carry different implications about the range of validity of (2), and about the correct way of extending it to more general situations. Even without answering it at all, however, we can still see the kind of difficulties that are going to face us. For if $f(x, p, t)$ is an average over something, then the left-hand side of (2) is also an average over this same something. So also, therefore, is the right-hand side if the equation is correct. But on the right-hand side we see the product of two f's; the product of two averages.

If you meditate about this for a moment, I think you will find it hard to avoid concluding that, if f is an average, then the right-hand side ought to contain the average of a product, not the product of the averages. These quantities are surely different; but we cannot say how different until we say what we are averaging over. *Until this ambiguity in the definition of Boltzmann's f is cleared up, we cannot assess the range of validity of* Eq. (2), *and we cannot say how it should be extended to more general problems*. Because of imprecise concepts, the theory reaches an impasse at the stage where it has barely scratched the surface of any real treatment of irreversible processes!

2.2. Method of GIBBS

For our second case history, we turn to the work of GIBBS. This was done some thirty years after the aforementioned work of BOLTZMANN, and the difficulties noted above, plus many others for which we do not have time here, were surely clear to GIBBS, who was extremely careful in matters of logic, detail, and definitions.

All important advances have their precursors, the full significance of which is realized only later; and the innovations of GIBBS were not entirely new. For example, considerations of the full phase space (Γ-space) appear already in the works of MAXWELL and BOLTZMANN; and GIBBS' canonical ensemble is clearly only a small step removed from the distribution laws of MAXWELL and BOLTZMANN. However, GIBBS applied these ideas in a way which was unprecedented; so much so that his work was almost totally rejected ten years later in the famous Ehrenfest review article [*5*], which has had a dominating influence on thought in statistical mechanics for fifty years. In this article, the methods of GIBBS are attacked repeatedly, and the physical superiority of BOLTZMANN'S approach is proclaimed over and over again. For example, GIBBS' canonical and grand canonical ensembles are dismissed as mere "analytical tricks", which do not solve the problem; but only enable GIBBS to *evade* what the authors consider to be real problems of the subject!

Since then, of course, the mathematical superiority of GIBBS' methods for calculating equilibrium thermodynamic properties has become firmly established; and so statistical mechanics has become a queer hybrid, in which the practical calculations are always based on the methods of GIBBS; while in the pedagogy virtually all one's attention is given to repeating the arguments of BOLTZMANN.

This hybrid nature — the attempt to graft together two quite incompatible philosophies — is nowhere more clearly shown than in the fact that the "official" commentary on GIBBS' work [*6*] devotes a major amount of space to discussion of ergodic theories. Now, it is a curious fact that if you study GIBBS' work, you will not find the word "ergodic" or the concept of ergodicity, at any point. Recalling that ergodic theorems, or hypotheses, had been actively discussed by other writers for over thirty years, and recalling GIBBS' extremely meticulous attention to detail, I think the only possible conclusion we can draw is that GIBBS simply *did not consider ergodicity as relevant to the foundations of the subject.* Of course, he was far too polite a man to say so openly; and so he made the point simply by developing his theory without making any use of it. Unfortunately, this tactic was too subtle to be appreciated by most readers; and the few who did notice it took it to be a defect in GIBBS' presentation, in need of correction by others.

This situation has had very unfortunate consequences, in that the work of GIBBS has been persistently misunderstood; and in particular, the full power and generality of the methods he introduced have not yet been recognized in any existing textbook. However, it is not a question of placing blame on anyone; for we can understand and sympathize with the position of everyone involved. I think that a historical study will convince you, as it has convinced me, that all of this is the more or less inevitable result of the fact that GIBBS did not live long enough to complete his work. The principle he had discovered was so completely new, and the method of thinking so completely different from what had gone before, that it was not possible to explain it fully, or to explore its consequences for irreversible phenomena, in the time that was granted to him.

GIBBS was in rapidly failing health at the time he wrote his work on statistical mechanics, and he lapsed into his final illness very soon after the manuscript was sent to the publisher. In studying his book, it is clear that it was never really finished; and we can locate very accurately the place where time and energy ran out on him. The first eleven chapters are written in his familiar style — extremely meticulous attention to detail, while unfolding a carefully thought out logical development. At Chapter 12, entitled, "On the Motion of Systems and Ensembles of Systems Through Long Periods of Time", we see an abrupt change of style; the treatment becomes sketchy, and amounts to little more than a random collection of observations, trying to state in words what he had not yet been able to reduce to equations. On pages 143—144 he tries to explain the methodology which led him to his canonical and grand canonical ensembles, as well as the ensemble canonical in the angular momenta which was presented in Chapter 4 but not applied to any problem [7]. However, he devotes only two sentences to this; and the principle he states is what we would recognize today as the principle of maximum entropy! To the best of my knowledge, this passage has never been noted or quoted by any other author (it is rather well hidden among discussions of other topics); and I discovered it myself only by accident, three years after I had written some papers [8] advocating this principle as a general foundation for statistical mechanics. This discovery convinced me that there was much more to the history of this subject than one finds in any textbook, and induced me to study it from the original sources; some of the resulting conclusions are being presented in this talk.

GIBBS' discussion of irreversibility in this chapter does not advance beyond pointing to a qualitative analogy with the stirring of colored ink in water; and this forms the basis for another of the EHRENFEST'S criticisms of his work. I think that, had GIBBS been granted a few more years of vigorous health, this would have been replaced by a simple and rigorous demonstration of the second law based on other ideas. For it turns out that all the clues necessary to point the way to this, and all the mathematical material needed for the proof, were already present in the first eleven chapters of his book; it requires only a little more physical reasoning to see that introduction of coarse-grained distributions does not advance our understanding of irreversibility and the second law, for the simple reason that the latter are experimentally observed *macroscopic* properties; and the fine-grained and coarse-grained distributions lead to just the same predictions for all macroscopic quantities. Thus, the difference between the fine-grained and coarse-grained H-functions has nothing to do with the experimentally observed entropy; it depends only on the particular way in which we choose to coarse-grain.

On the other hand, the variational (maximum entropy) property noted by GIBBS does lead us immediately to a proof, not only of the second law, but of an extension of the second law to nonequilibrium states. I have recently pointed this out [*9*] and supplied the very simple proof, which I think is just the argument GIBBS would have given if he had been able to complete his work. However, this is not the main point I wish to discuss tonight, so let us turn back to other topics.

In defense of the EHRENFEST'S position, it has to be admitted that, through no fault of his own, GIBBS did fail to present any clear description of the motivation behind his work. I believe that it was virtually impossible to understand what GIBB'S methods amounted to, *and therefore how great was their generality and range of validity*, until the appearance of SHANNON'S work on Information Theory, in our own time [*10*]. Finally, until recently the situation in probability theory itself, which was in a high-pressure phase completely dominated by the frequency theory, which only sneers and attacks on the theories of LAPLACE and JEFFREYS, has made it impossible even to discuss, much less publish, the viewpoint and approach which I believe has now solved these problems.

Now, in order to lend a little more substance to these remarks, let's examine some equations, the net result of GIBBS' work. Considering a closed system (i.e., no particles enter or leave), the thermodynamic properties are to be calculated from the Hamiltonian $H(q_i, p_i)$ as follows. First, we define the *partition function*

$$Z(\beta, V) \equiv \int \exp(-\beta H)\, dq_1\, dp_1 \ldots dq_n\, dp_n, \tag{5}$$

where we integrate over all the accessible phase space, and the dependence on the volume V arises because the range of integration over the coordinates q_i depends on V. If we succeed in evaluating this function, then all thermodynamic properties are known; for the energy function (which determines the thermal properties) is given by

$$U = -\frac{\partial}{\partial \beta} \log Z \tag{6}$$

in which we interpret β as $(kT)^{-1}$, where k is BOLTZMANN'S constant and T the KELVIN temperature; and the equation of state is

$$P = \frac{1}{\beta} \frac{\partial}{\partial V} \log Z. \tag{7}$$

Now, isn't this a beautifully simple and neat prescription? For the first time in what has always been a rather messy subject, one had a glimpse of the kind of formal elegance that we have in mechanics, where a single equation (HAMILTON'S principle) summarizes everything that needs to be said. Of all the founders of statistical mechanics, only GIBBS gives us this formal simplicity, generality, and as it turned out, a technique for practical calculation which the labors of another sixty years have not been able to improve on. The transition to quantum statistics took place so quietly and uneventfully because it consisted simply in the replacement of the integral in (5) by the corresponding discrete sum; and nothing else in the formalism was altered.

In the history of science, whenever a field has reached such a stage, in which thousands of separate details can be summarized by, and deduced from, a single formal rule — then an extremely important synthesis has been accomplished. Furthermore, by understanding the basis of this rule it has always been possible to extend its application far beyond the original set of facts for which it was designed. And yet, this did not happen in the case of GIBBS' formal rule. With only a few exceptions, writers on statistical mechanics since GIBBS have tried to snatch away this formal elegance by grafting GIBBS' method onto the substrate of BOLTZMANN'S ideas, for which GIBBS himself had no need. However, a few, including TOLMAN and SCHRÖDINGER, have seen GIBBS' work in a different light — as something that can stand by itself without having to lean on unproved ergodic hypotheses, intricate but arbitrarily defined cells in phase space, Z-stars, and the like. Thus, while a detailed study will show that there are as many different opinions as to the reason for GIBBS' rules as there are writers on the subject, a more coarse-grained view shows that these writers are split into two basic camps; those who hold that the ultimate justification of GIBBS' rules must be found in ergodic theorems; and those who hold that a principle for assigning *a priori* probabilities will provide a sufficient justification. Basically, the confusion that still exists in this field arises from the fact that, while the *mathematical content* of GIBBS' formalism can be set forth in a few lines, as we have just seen, the *conceptual basis* underlying it has never been agreed upon.

Now, while GIBBS' formalism has a great generality — in particular, it holds equally well for gas and condensed phases, while BOLTZMANN'S results apply only to dilute gases — it nevertheless fails to give us many things that BOLTZMANN'S "collision equation" does yield, however imperfectly. For BOLTZMANN'S equation can be applied to irreversible processes; and it gives definite theoretical expressions for transport coefficients (viscosity, diffusion, heat conductivity), while GIBBS' rules refer only to thermal equilibrium, and one has not seen how to extend them beyond that domain. Furthermore, in spite of all my carping about the imprecision of BOLTZMANN'S equation, the fact remains that it has been very successful in giving good numerical values for these transport coefficients; and it does so even for fairly dense gases, where we really have no right to expect such success. So, my adulation of Gibbs must be carried to the point of rejecting BOLTZMANN'S work; it appears that we need both approaches!

All right. I have now posed the problem as it appeared to me a number of years ago. Can't we learn how to combine the best features of both approaches, into a new theory that retains the unity and formal simplicity of GIBBS' work with the ability to describe irreversible processes (hopefully, a *better* ability) of BOLTZMANN'S work? This question must have occurred to almost every physicist who has made a serious study of statistical mechanics, for the past sixty years. And yet, it has seemed to many a hopelessly difficult task; or even an impossible one. For example, at the 1956 International Congress on Theoretical Physics, L. VAN HOVE [*11*] remarked, "In contrast to the case of thermodynamical equilibrium, no general set of equations is known to describe the behavior of many-particle systems whenever their state is different from the equilibrium state and, in view of the unlimited diversity of possible nonequilibrium situations, the existence of such a set of equations seems rather doubtful".

Now, while I hesitate to say so at a symposium devoted to Philosophy of Science, the injection of philosophical considerations into science has usually proved fruitless, in the sense that it does not, of itself, lead to any advances in the science. But there is one extremely important exception to this; and it is in exactly the situation now before us. At the stage in development of a theory where we already have a formalism successful in one domain, and we are trying to extend it to a wider one, some kind of philosophy about what the formalism "means" is absolutely essential to provide us with a sense of direction. And it need not even be a "true" philosophy — whatever that may mean — for its real justification will not lie in whether it is "true", but in whether it does point the way to a successful extension of the theory.

In the construction of theories, a philosophy plays somewhat the same role as scaffolding does in the construction of buildings; you need it desperately at a certain phase of the operation, but when the construction is completed you can remove if it you wish; and the structure will still stand of its own accord. This analogy is imperfect, however, because in the case of theories, the scaffolding is rarely ugly, and many will wish to retain it as an integral part of the final structure. At the opposite extreme to this conservative attitude stands the radical positivist, who in his zeal to remove every trace of scaffolding, also tears down part of the building. Almost always, the wisest course will lie somewhere between these extremes.

The point which I am trying to make, in this rather cryptic way, is just the one which we have already noted in the attempt to evaluate and extend BOLTZMANN's collision equation. Different philosophies of what that equation means carry different implications as to its range of validity, and the correct way of extending it. And we are now at just the same impasse with regard to GIBBS' equations; *because their conceptual basis has not been precisely defined, the theory dissolves in ambiguities* which have prevented us, for sixty years, from extending to to new domains.

2.3. Conceptual Problems of the Ensemble

The fact that two different camps exist, with diametrically opposed views as to the justification of GIBBS' methods, is simply the reflection of two diametrically opposed philosophies about the real meaning of the GIBBS ensemble; and this in turn arises from two different philosophies about the meaning of *any* probability distribution. Thus, the foundations of probability theory itself are involved in the problem of extending GIBBS' methods.

Statistical mechanics has always been troubled with questions concerning the relation between the ensemble and the individual system, even apart from possible extensions to nonequilibrium cases. In the theory, we calculate numbers to compare with experiment by taking ensemble *averages*; that is what we are doing in Eqs. (6) and (7). And yet, our experiments to check these predictions are not performed on ensembles; they are performed on the one *individual* system that exists in the laboratory. Nevertheless, we find that the predictions are verified accurately; a rather astonishing result, but one without which we would have little interest in ensembles. For if it were necessary to repeat a thermodynamic measurement 1,000 times and average the results before any regularities (laws of thermodynamics) began to appear, both thermodynamics and statistical mechanics would be virtually useless to us; and they would not appear in our physics curriculum. Thus, it appears that a major problem is to explain why GIBBS' rules work in practice; and not only why they work so well, but why they work at all!

We can make this dilemma appear still worse by noting that the relation between the ensemble and the individual system is usually described by supposing that the individual system can be regarded as having been drawn "at random" from the ensemble. I personally have never been able to comprehend what "at random" means; for I ask myself: What is the criterion, what is the test, by which we could decide whether it was or was not really "random"? Does it make sense to ask whether it was *exactly* random, or *approximately* random? — and neither the literature nor my introspection give me any answer. However, even without understanding this point, the real difficulty is obvious; for the *same* individual system may surely, and with equal justice, be regarded as having been drawn "at random" from any one of an infinite number of *different* ensembles! But the measured properties of an individual system depend on the *state of the system*; and not on which ensemble you or I regard it as having been "drawn from". How, then is it possible that ensemble averages coincide with experimental values?

The two different philosophical camps try to extricate themselves from this dilemma in two entirely different ways. The "ergodic" camp, of course, is composed of those who believe that a probability distribution describes an objectively real physical situation; that it stands for an assertion about experimentally measurable *frequencies*; that it is therefore either correct or incorrect; and that this can, in principle, be decided by performing "random experiments". They note that what we measure in any experiment is necessarily a time average over a time that is long on the atomic scale of things; and so the success of GIBBS' methods will be accounted for if we can prove, from the microscopic equations of motion, that the *time average* for an individual system is equal to the *ensemble average* over the particular ensembles given by GIBBS.

This viewpoint has much to recommend it. In the first place, physicists have a natural tendency to believe that, since the observed properties of matter "in the large" are simply the resultant of its properties "in the small" multiplied many times over, it ought to be possible to obtain the macroscopic behavior by strict logical deduction from the microscopic laws of physics; and the "ergodic" approach gives promise of being able to do this. Secondly, while the necessary theorems have not been established rigorously and universally, the work done on this problem thus far has made it highly plausible that, in a system interacting with a large heat bath, the *frequencies* with which various microscopic conditions are realized in the long run are indeed given correctly by the GIBBS canonical ensemble. This has been rendered so extremely plausible that I think no reasonable person can seriously doubt that it is true, although we cannot rule out the possibility of occasional "pathological" exceptions. Thus the "ergodic" school of thought has, in my opinion, very nearly succeeded in its aim of establishing equality of time averages and ensemble averages *for the particular case of Gibbs' canonical ensemble*; and in the following I am simply going to grant, for the sake of the argument, that this program has succeeded entirely.

Nevertheless, the "ergodic" school of thought still faces a fundamental difficulty; and one that was first pointed out by BOLTZMANN himself, and stressed in the EHRENFEST review article. Curiously, there exists to this day a group of workers in Europe who refuse to recognize the seriousness of this difficulty, and deny that it invalidates their approach. The difficulty is that, even if one had succeeded in proving these ergodic theorems rigorously and universally, the result would have been established only for time averages *over infinite times*; whereas the experiments which verify GIBBS' rules measure time averages only over finite times. Thus, a further mathematical demonstration would in any event be necessary, to show that these finite time averages have sufficiently approximated their limits for infinite times.

Now we can give simple and general counter-examples proving that such an additional demonstration *cannot* be given; and indeed that any macroscopic system, given a time millions of times the age of the universe, still could not "sample" more than an infinitesimal fraction of all the microscopic states which have high probability in the canonical emsemble; and thus any assertion about the *frequencies* with which different microscopic states are realized in an individual system, is completely devoid of operational meaning.

The easiest way of seeing this is just to note that, if a macroscopic system could sample all microscopic states in the time in which measurements are made, so that the measured time averages would be equal to ensemble averages, then the measured values would necessarily always be the *equilibrium* values; we would not even know about irreversible processes! *The fact that we can measure the rate of an irreversible process already proves that the time required for a representative sampling of microstates must be much longer than the time required to make our measurements.* Thus, any purported proof that time averages over the finite times involved in actual measurements are equal to canonical ensemble averages would, far from justifying statistical mechanics, stand in clear conflict with the very experimental facts about irreversibility that we are trying to account for by extending GIBBS' methods!

The thing which has to be explained is, not that ensemble averages are equal to time averages; but the much stronger statement that ensemble averages are equal to experimental values. The most that ergodic theorems could possibly establish is that ensemble averages are equal to time averages over infinite time, and so the "ergodic" approach cannot even justify equilibrium statistical mechanics without contradicting experimental facts. Obviously, such an approach cannot be extended to irreversible processes where, in order for ensemble theory to be of any use, the ensemble averages must still be equal to experimental values; but the very phenomena to be explained consist of the fact that these are *not* equal to time averages.

The above line of reasoning convinced me, ten years ago, that further advances in the basic formulation of statistical mechanics cannot be made within the framework of the "ergodic" viewpoint; and, rightly or wrongly, it seemed equally clear to me that the really fundamental trouble which was preventing further advances, both in statistical mechanics and in the field of statistics in general, was this dogmatic, single-minded insistence on the frequency theory of probability which had dominated the field for so many years. At that time, virtually every writer on probability theory felt impelled to insert an introductory paragraph or two, expressing his denunciation and total rejection of the so-called "subjective" interpretation of probability, as advocated by LAPLACE, DE MORGAN, POINCARÉ, KEYNES, and JEFFREYS; and this was done, invariably, without any attempt to understand the arguments and results which these people — particularly LAPLACE and JEFFREYS — had advanced. The situation was, psychologically, exactly like the one which has dominated American Politics since about 1930; the Republicans continually analyze the statements of Democrats and issue counter-arguments, which the Democrats contemptuously dismiss without any attempt to understand them or answer them.

On the other hand, I had taken the trouble to read all of JEFFREYS' work, and much of LAPLACE'S, on probability theory; and was unable to find any of the terrible things about which the "frequentist" writers had warned us. On the philosophical side I found their arguments to be, far from irresponsible and useless, so eminently sound and reasonable that I could not imagine any sane person disputing them. On the mathematical side, I found that in problems of statistical estimation and hypothesis testing, any problem for which the "frequentist" offered any solution at all was also solved with ease by the methods of LAPLACE and JEFFREYS; and their results were either the same or demonstrably superior to the ones found by the frequentists. Furthermore, the methods of LAPLACE and JEFFREYS (which were, of course, based on BAYES' theorem as the fundamental tool of statistics) were applied with equal ease to many problems which, according to the frequentist, did not belong to the field of probability theory at all; and they still yielded perfectly reasonable, and scientifically useful, results!

I don't want to dwell at length on the situation in probability theory, because time is running short and a rather large exposition of this,with full mathematical details, is being readied for publication elsewhere. But let me just mention one example of what one finds if he takes the trouble to go beyond polemics and study the mathematical facts of the matter.

In problems of interval estimation of unknown parameters, the frequentist has rejected the method of LAPLACE and JEFFREYS, on grounds that I can only describe as ideological, and has advocated vigorously the method of confidence intervals. Now it is a matter of straightforward mathematics to show that, whenever the frequentist's "estimator" is not a sufficient statistic (in the terminology of FISHER), there is always a class of possible samples for which the method of confidence intervals leads to absurd or dangerously misleading results, in the sense that it yields a wrong answer far more frequently (or, if one prefers, with far higher probability) than one would suppose from the stated confidence level. The confidence interval can, in some cases, contradict what can be proved on strict deductive reasoning from the observed sample. One can even invent problems, which are not at all unrealistic, in which the probability of this happening is greater than the stated confidence level!

This is something which, to the best of my knowledge, you cannot find mentioned in any of the "orthodox" statistical literature; and I shudder to think of some of the possible consequences, if important decisions are being made on the basis of confidence interval analyses. The method of LAPLACE and JEFFREYS is demonstrably free from this defect; it cannot contradict deductive reasoning and, in the case of the aforementioned "bad" class of samples, it automatically detects them and yields a wider interval, so that the probability of a correct decision remains equal to the stated value. Once one is aware of such facts, the arguments advanced against the method of LAPLACE and JEFFREYS and in favor of confidence intervals (i.e. that it is meaningless to speak of the probability that θ lies in a certain interval, because θ is not a "random variable," but only an unknown constant) appear very much like those of the 17th century scholar who claimed his theology had proved there could be *no* moons on Jupiter, and steadfastly refused to look through GALILEO's telescope.

Since the reasoning by which the "frequentist" has rejected LAPLACE's methods is so patently unsound, and since attempts to extend, or even justify, GIBBS' methods in terms of the frequency theory of probability have met with an impasse, it would appear that we ought to explore the possibilities of applying LAPLACE's "subjective" theory of probability to this problem. At any rate, to reject this procedure without bothering to explore its potentialities, is hardly what we mean by a "scientific" attitude! So, I undertook to think through statistical mechanics all over again, using the concept of "subjective" probability.

It became clear, very quickly, that to do this makes all the unsolved problems of the theory appear in a very different light; and possibilities for extension of GIBBS' methods are seen in entirely different directions. Once we clearly and explicitly free ourselves from the delusion that an ensemble describes an "objectively real" physical situation, and recognize that it describes only a certain *state of knowledge*, then it is clear that, in the case of irreversible processes, the knowledge which we have is of a different nature than in the case of equilibrium. We can then see the problem as one which cannot even be formulated in terms of the frequency theory of probability. It is simply this: *What probability assignment to microstates correctly describes the state of knowledge which we have, in practice, about a nonequilibrium state?* Such a question just doesn't make sense in terms of the frequency theory; but, thanks to the work of GIBBS and SHANNON, I believe that it makes extremely good sense, and in fact has a very general and mathematically unambiguous solution in terms of subjective probabilities.

3. The General Maximum-Entropy Formalism

If we accept SHANNON's interpretation (which can be justified by other mathematical arguments entirely independent of the ones given by SHANNON) that the quantity

$$H = -\sum_i p_i \log p_i \tag{8}$$

is an "information measure" for any probability distribution p_i; i.e. that it measures the "amount of uncertainty" as to the true value of i, then an ancient principle of wisdom — that one ought to acknowledge frankly the full extent of his ignorance — tells us that the distribution that maximizes H subject to constraints which represent whatever information we have, provides the most honest description of what we know. The probability is, by this process, "spread out" as widely as possible without contradicting the available information.

But recognition of this simple principle suddenly makes all the maximum-minimum properties given by GIBBS in his Chapter XI — what I believe to be the climax of GIBBS' work, and just the place where time and energy ran out on him — acquire a much deeper meaning. If we specify the expectation value of the energy, this principle uniquely determines GIBBS' canonical ensemble. If we specify the expectations of energy and mole numbers, it uniquely determines GIBBS' grand canonical ensemble [8]. If we specify the expectations of energy and angular momentum, it uniquely determines GIBBS' rotational ensemble [7]. Thus, all the results of GIBBS on statistical mechanics follow immediately from the principle of maximum entropy; and their derivation is astonishingly short and simple compared to the arguments usually found in textbooks.

But the generalization of GIBBS' formalism to nonequilibrium problems also follows immediately (although I have to confess that I spent six years trying to do this by introducing new and more complicated principles, before I finally saw how simple the problem was). For this principle in no way depends on the physical meaning of the quantities we specify; there is nothing unique about energy, mole numbers, or angular momentum. If we grant that it represents a valid method of reasoning at all, then we must also grant that it applies equally well to *any physical quantity whatsoever*. So, let us jump immediately, in view of the time, to the most sweeping generalization of GIBBS' formalism.

We have a number of physical quantities about which we have some experimental information. Let them be represented by the Heisenberg operators $F_1(x, t)$, $F_2(x, t)$, ... $F_m(x, t)$. In general they will depend on the position x and, through the equations of motion, on the time t. For example, F_1 might be the particle density, F_2 the density of kinetic energy, F_3 the "mass velocity" of the fluid, F_4 the (yz)-component of the stress tensor, F_5 the intensity of magnetization, ..., and so on; whatever information of this type is available, represents our definition of the nonequilibrium state.

Now we wish to construct a density matrix ϱ which incorporates all this information. When I say that a density matrix "contains" certain information, I mean by this simply that, if we apply the usual rule for prediction; i.e. calculate the expectation values

$$\langle F_k(x, t)\rangle = \mathrm{Tr}[\varrho\, F_k(x, t)] \tag{9}$$

we must be able to recover this information from the density matrix. Thus, the mathematical constraints on the problem are that the expectation values (9) must agree with the experimental information:

$$f_k(x, t) = \mathrm{Tr}[\varrho\, F_k(x, t)], \qquad x, t \text{ in } R_k \tag{10}$$

where $f_k(x, t)$ represent the experimental values, and R_k is the space-time region in which we have information about f_k; in general it may be different for different k. Subject to these constraints, we are to maximize the "information entropy"

$$S_I = -\mathrm{Tr}(\varrho \log \varrho) \tag{11}$$

which is the appropriate generalization of (8), as found many years ago by VON NEUMANN. The solution of this variational problem is:

$$\varrho = \frac{1}{Z} \exp\left\{\sum_{k=1}^{m} \int_{R_k} d^3x\, dt\, \lambda_k(x, t)\, F_k(x, t)\right\} \tag{12}$$

where the $\lambda_k(x, t)$ are a set of real functions to be determined presently (they arise mathematically as Lagrange multipliers in solving the variational problem with constraints), and for normalization the partition function of GIBBS has been generalized to the *partition functional:*

$$Z[\lambda_k(x,t)] \equiv \operatorname{Tr} \exp\left\{\sum_{k=1}^{m} \int_{R_k} d^3x\, dt\, \lambda_k(x,t)\, F_k(x,t)\right\}. \tag{13}$$

The $\lambda_k(x,t)$ are now to be found from the conditions (10), which reduce to

$$f_k(x,t) = \frac{\delta}{\delta \lambda_k(x,t)} \log Z \tag{14}$$

which is a generalization of GIBBS' equation (6); where δ denotes the functional derivative. Mathematical analysis shows that (14) is just sufficient to determine uniquely the integrals in the exponent of (12); it does not necessarily determine the functions $\lambda_k(x,t)$, but it does determine the only property of those functions which is needed in the theory; a very interesting example of mathematical economy.

The density matrix having been thus found, prediction of any other quantity $J(x,t)$ in its space-time dependence is then found by applying the usual rule:

$$\langle J(x,t)\rangle = \operatorname{Tr}[\varrho J(x,t)]. \tag{15}$$

In Eqs. (12) to (15) we have the generalization of GIBBS' algorithm to arbitrary nonequilibrium problems. From this point on, it is simply a question of mathematics to apply the theory to any problem you wish.

Of course, it requires a great deal of nontrivial mathematics to carry out these steps explicitly for any nontrivial problem! If GIBBS' original formalism was somewhat deceptive, in that its formal simplicity conceals an enormous amount of intricate detail, the same is true with a vengeance for this generalization. Nevertheless, it is still only mathematics; and if it were important enough to get a certain result, one could always hire a building full of mathematicians and computers to grind it out; there are no further questions of principle to worry about.

For the past three years, my students and I have been exploring these mathematical problems, and we have a large mass of results that will be reported in due course. Without going into further details, let me just say that all the previously known results in theory of irreversible processes can be derived easily from this algorithm. Dissipative effects such as viscosity, diffusion, heat conductivity are obtained by direct quadratures using (15), with no need for the forward integration and coarse-graining operations characteristic of previous treatments. For static transport coefficients we obtain formulas essentially equivalent to those of KUBO; we can exhibit certain ensembles for which KUBO'S results, originally obtained by perturbation theory, are in fact exact.

Because we are freed from the need for time-smoothing and other coarse-graining operations, the theory is no longer restricted to the quasi-stationary, long-wavelength limit. It gives, with equal ease, general formulas for such things as ultrasonic attenuation and for nonlinear effects, such as those due to extremely large temperature or concentration gradients, for which previously no unambiguous theory existed. Because of these results, I feel quite confident that we are on the right track, and that this generalization will prove to be the final form of nonequilibrium statistical mechanics.

Let me close with a couple of philosophical remarks, relating this development to things I mentioned earlier in this talk. In seeking to extend a theory to new domains, some kind of philosophy about what the theory "means" is absolutely essential. The philosophy which led me to this generalization was, as already indicated, my conviction that the "subjective" theory of probability has been subjected to grossly unfair attacks from people who have never made the slightest attempt to examine its potentialities; and that if one does take the trouble to rise above ideology and study the facts, he will find that "subjective" probability is not only perfectly sound philosophically; it is a far more powerful tool for solving practical problems than the frequency theory. I am, moreover, not alone in thinking this, as those familiar with the rise of the "neo-Bayesian" school of thought in statistics are well aware.

Nevertheless, that philosophy of mine was only scaffolding, which served the purpose of telling me in what *specific* way the formalism of GIBBS was to be generalized. Once a philosophy has led to a definite, unambiguous mathematical formalism by which practical calculations may be carried out, then the issue is no longer one of philosophy; but of fact. The formalism either will or will not prove adequate in practice; and it will be judged, quite properly, not by the philosophy which led to it, but by the results which its gives. If you do not like my philosophy, but you find that the formalism, nevertheless, does give useful results, then I am quite sure that you will be able to invent some *other* philosophy by which that formalism can be justified! And, perhaps, that other philosophy will lead to still further generalizations and extensions, to which my own philosophy makes me blind. That is, after all, just the process by which all progress in theoretical physics has been made.

REFERENCES

[1] EPSTEIN, P. S.: Textbook of thermodynamics, p. 27—34. New York: John Wiley & Sons, Inc. 1937.

[2] BELL, E. T.: Men of mathematics, p. 546. New York: Dover Publ. Inc. 1937.

[3] See, for example: Niels Bohr and the development of physics (W. PAULI, ed.), p. 17—28, and footnote, p. 76. New York: Pergamon Press 1955; Observation and interpretation (S. KÖRNER, ed.), p. 41—45. New York: Academic

Press, Inc. 1957; W. Heisenberg, Physics and philosophy, p. 128—146. New York: Harper & Brothers, Publ. 1958; N. R. Hanson, Am. J. Phys. **27**, 1 (1959); Quanta and reality (D. Edge, ed.) p. 85—93. Larchmont (New York): Am. Research Council 1962.

[*4*] Dresden, M.: Revs. Mod. Phys. **33**, 265 (1961).

[*5*] Ehrenfest, P. and T.: Encykl. Math. Wiss. 1912. English translation by M. J. Moravcsik, The conceptual foundations of the statistical approach in mechanics. Ithaca (N. Y.): Cornell University Press 1959.

[*6*] Gibbs, J. W.: Collected works and commentary, vol. II (A. Haas, ed.), p. 461—488. Yale University Press New Haven (Conn.): 1936.

[*7*] A successful application of Gibbs' rotationally canonical ensemble to the theory of gyromagnetic effects has since been given: S. P. Heims and E. T. Jaynes. Revs. Mod. Phys. **34**, 143 (1962).

[*8*] Jaynes, E. T.: Phys. Rev. **106**, 620; **108**, 171 (1957).

[*9*] — Chapter 4 of Statistical physics (1962 Brandeis Lectures) (K. W. Ford, ed.). New York: W. A. Benjamin, Inc. 1963; Gibbs vs Boltzmann entropies, Am. J. Phys. **33**, 391 (1965).

[*10*] Shannon, C. E., and W. Weaver: The mathematical theory of communication. Urbana (Ill.): University of Illinois Press 1949.

[*11*] Hove, L. van: Revs. Mod. Phys. **29**, 200 (1957).

Chapter 7

General Covariance in Electromagnetism

E. J. Post
Microwave Physics Laboratory
Air Force Cambridge Research Laboratories
Bedford, Massachusetts

1. Introduction

Although the principle of general covariance is almost fifty years old, physicists are not yet agreed on its meaning and value as a tool in physics. The arguments against it range somewhat as follows: "Covariance is a mathematical artifice — no physical meaning can be attached to it. Every equation can be expressed covariantly. How can a covariant form be right or wrong from a physical point of view? — covariance is only a method of expressing given physical relations in a more pleasing and elegant mathematical form." Representative of the more prevalent opinion are the arguments holding out partial acceptance. These are roughly the following: "It is clear that Lorentz covariance plays an important role in physics. There is, however, no adequate proof or evidence that extended forms of covariance have any explicit meaning. The physical implications of general relativity are still too uncertain to justify an acceptance of general covariance as a well-established physical principle; however, a general covariance in the sense of a natural extension of LAGRANGE's method in point mechanics is acceptable."

These argued opinions are reminiscent of an early polemic between KRETSCHMANN and EINSTEIN, which appeared in the literature almost fifty years ago [*1*, *2*]. After EINSTEIN had introduced the principle of general covariance, KRETSCHMANN made the rather startling statement that the principle of general covariance was purely formal in character and therefore could not generate any necessary physical results. This statement was in a way sensational: even if KRETSCHMANN was not at that time prepared to accept general relativity, it was at least true that EINSTEIN's famous $E = mc^2$ law, which had been derived from the more restricted requirement of Lorentz covariance, was by then more than ten years old. And there was then already definite indication that this law had more than just a formal meaning, although the evidence in those days was by no means so overwhelming as today.

The resolution of the Kretschmann-Einstein controversy brought out two important facts, one concerning the physical nature of the things that may be expected from general covariance, and the other concerning the physically meaningful implementation of general covariance.

Let us first discuss the nature of the physical results that ensue from an application of general covariance. This is best illustrated by comparing them with the physical results yielded by the more restricted requirement of Lorentz covariance. The $E=mc^2$ law, for instance, has in it a parameter c^2 that originates in the specific character of the Lorentz transformations. It thus seems that new results ensuing from a requirement of Lorentz covariance are of a specific quantitative character because the Lorentz transformations have specific quantitative parameters and apply to specific physical situations (freespace).

The transformations to which we refer when we speak of general covariance are usually not specified in detail, although they certainly contain the Lorentz group as a subset. The set of general transformations might be called amorphous in contrast to the specific subset known as the Lorentz group. The physical consequences that one may expect from general covariance are therefore of a more qualitative nature.

It would be somewhat shortsighted if we discarded results just because of their qualitative nature. A consistency argument, for instance, can be of a qualitative nature. And then, it is sometimes possible to derive quantitative statements from a qualitative statement by forcing an asymptotic comparison with an accepted specific result. The gravitational field equations of general relativity are constructed in that manner.

We thus concur with EINSTEIN that the amorphous nonspecific nature of the transformations does not justify a conclusion that the principle is physically empty. For physical results the emphasis should be on the appropriate applications of the principle rather than on premature philosophic digressions on the nature of the principle.

2. The Implementation of General Covariance

How do we apply general covariance, and when? Is it possible that a given law of physics can be rendered in a generally covariant form in more than one way? These are questions that concern the implementation of general covariance in physics. Imposing the requirement of general covariance is no more no less than stating an epistemological objective. Only the implementation can yield physical results. The characteristic features of covariance implementation can be best illustrated by some examples.

No other set of equations in physics has been rendered in more different covariant forms than the Maxwell equations. The Maxwell equations have appeared in the literature in the three-dimensional vector form, space-time tensor form, bra-and-ket form, and second- and first-rank spinor forms. The vector and tensor forms can be split further into tensor or vector forms with real or complex components. Covariance has been discussed with respect to spatial rotations, affine transformations, conformal transformations, curvilinear transformations, the Lorentz group, and the set of general space-time transformations and its subset that preserves the space-time volume element. Only the rotational and the Lorentz invariant forms are in general use.

The basic question for any transformation of independent variables of a set of equations is the question of what we need to assume about the transformation of the dependent variables. Mathematically we are free to keep the dependent variables the way they are. The transformation of independent variables is then solely for the purpose of obtaining a mathematically different or simpler form. From this mathematically simpler form we would in general not be able to reconstruct the original physical problem unless we were given the transformation that led to this simpler form.

In physics (and also in geometry) it is customary to prefer a method of transformation such that the physical information contained in the original formulation is preserved. This usually requires a transformation of both the dependent and independent variables, the coordinate transformation inducing a transformation of the dependent variables such that the "form" of the equation is preserved. "Form" of the equations means an aggregation of structural characteristics that are translatable into physical information. For instance, the Maxwell, Schrödinger, and Dirac equations have structural distinctions that relate them to different physical situations.

The main problem in implementing covariance thus becomes a search for the dependent variables whose induced transformations preserve the physical information contained in the equations. The method of transformation must relate observations made in one frame of reference to observations made in another frame of reference. Problems of this kind, as is well known, have led to the concepts of vectors, tensors, and spinors as prototypes of typical dependent variables. Along with such dependent variables as vectors and tensors that obey homogeneous rules of (induced) transformation, there are also some that do not. There is a very characteristic physical distinction associated with these different rules of transformation. A field described by a dependent variable obeying an inhomogeneous transformation rule can be generated by a change of coordinates (for example, inertial force) whereas a vector that vanishes in one system will vanish in any system of coordinates.

In implementing covariance, what determines the group of transformation? Why should we aim for the widest possible group, as implied by the principle of general covariance?

It must be remembered that groups are frequently used to characterize the symmetries of a physical medium. The crystallographic point groups, for instance, are used to classify crystals. The Lorentz group characterizes such physical properties as those of freespace. When we deal with the application of a physical law to a medium of given symmetry it is imperative that the form of that physical law permit the symmetry operations associated with the given medium. In other words, the invariance group of that physical law should at least be a covering group of the symmetry group of the medium. Hence, a general applicability of the law to arbitrary media will require the widest possible invariance group so as to accommodate the greatest variety of different physical media.

The criterion of a basic physical law is that it obey form invariance under the widest possible group — in this context, the set of general space-time transformations. This assures a functional separation between the relations expressing the law and the relations describing the specific properties of the medium under consideration. General covariance was never meant to be indiscriminately imposed on the relations describing the physical properties of a medium.

The preceding remarks explain why the LORENTZ and rotational invariant forms are the most frequently used: a wider invariance is unnecessary for isotropic media.

In retrospect, the many intermediate invariance studies of, for instance, the Maxwell equations (rotational, LORENTZ, conformal invariance, etc.) ultimately culminated in the "natural" general invariance; each of these intermediate invariance studies in some way or another displays the symmetry of a medium. The latter fact has not always been explicitly mentioned, because not every transformation subgroup necessarily represents a realistic physical medium.

On the question of uniqueness of general covariance, it is true that form invariance for general space-time transformations can be obtained in many different ways. But how do we select the physically most meaningful, that is, the natural invariant form? The guiding principle is to aim at a general invariant from that maximizes the physical situations to which the law applies. The key to the procedure lies in choosing the proper transformation behavior of the dependent variables, as will be illustrated in this paper.

3. Inconsistencies Solvable by General Covariance

The second set of Maxwell equations

$$\begin{aligned} \operatorname{curl} \boldsymbol{H} &= \dot{\boldsymbol{D}} + \mathbf{s} \\ \operatorname{div} \boldsymbol{D} &= \varrho \end{aligned} \tag{1}$$

can be put into a space-time invariant form. Some textbooks give the following generally covariant form of these equations:

$$g^{-\frac{1}{2}} \partial_\nu (g^{\frac{1}{2}} G^{\lambda\nu}) = c^\lambda, \qquad \partial_\nu = \frac{\partial}{\partial x^\nu}, \tag{2a}$$

where $G^{\lambda\nu}$ is an antisymmetric tensor (also called bivector or six-vector) taking the place of the two spatial vectors $\boldsymbol{D}$ and $\boldsymbol{H}$, and c^λ is the four-vector of charge and current. The factor $g^{\frac{1}{2}}$ is the square root of the absolute value of the determinant of the metric and ensures the general covariance of (2a).

Other textbooks (much fewer though) give the following equation as the covariant equivalent of (1)

$$\partial_\nu \mathfrak{G}^{\lambda\nu} = \mathfrak{c}^\lambda, \tag{2b}$$

where

$$\begin{aligned} \mathfrak{G}^{\lambda\nu} &= g^{\frac{1}{2}} G^{\lambda\nu}; \\ \mathfrak{c}^\lambda &= g^{\frac{1}{2}} c^\lambda. \end{aligned} \tag{3}$$

The quantity $\mathfrak{G}$ in (2b) has the properties of a density of weight $+1$, and so has $\mathfrak{c}$. The question now is, which of the two quantities $G, \mathfrak{G}$ represents the $\boldsymbol{H}, \boldsymbol{D}$ field and which of the two quantities, c, $\mathfrak{c}$ represents the charge density ϱ and current density $\mathbf{s}$?

The problem can be resolved in a number of ways. First, it should be noted that the vector $\boldsymbol{D}$ as a charge displacement per unit area clearly has the properties of a spatial density, which would strongly bias us in favor of $\mathfrak{G}$ because a space-time density will reduce to a spatial density if the time is not affected by the transformation. The vector $\boldsymbol{H}$, however, has the character of a covariant vector, with the important distinction that it is not affected by an inversion of coordinates. Such a (covariant) pseudo vector can be represented by a bivector density by means of the so-called Levi-Civita or Ricci tensor density, also known as the alternating unit tensor. Hence we have now completed the argument that the spatial properties of $\boldsymbol{D}$ and $\boldsymbol{H}$ support the idea that $\mathfrak{G}^{\lambda\nu}$, rather than $G^{\lambda\nu}$, is the actual field. Similar arguments support $\mathfrak{c}^\lambda$ rather than c^λ as the four-vector representing charge and current density.

The choice between (2a) and (2b) can also be discussed on the basis of a consistency argument. Consider a domain of space where charge and current vanish, which means that the righthand member in both

Eqs. (2a) and (2b) is zero. We would expect this observation to be independent of the state of motion of the observer, including accelerated motion. Accelerated motion is always associated with a nonlinear transformation of time and space coordinates; hence, a generally covariant formulation is indicated for discussing the problem. The formulation (2b) clearly preserves the zero righthand member, whereas an expansion of (2a) gives

$$\partial_\nu G^{\lambda\nu} + \tfrac{1}{2} G^{\lambda\nu} \partial_\nu \ln g = 0, \tag{4}$$

where the extra term depending on g requires a physical interpretation. If G really represents the $\boldsymbol{D}$, $\boldsymbol{H}$ field, then it would seem that an accelerated motion could generate an equivalent of charge and current. The time component ($\lambda = 0$) of Eq. (4) could then lead to an expression

$$\operatorname{div} \boldsymbol{D} \neq 0. \tag{5}$$

The extra term in (4) is of course produced by the covariant derivative, and it can be argued that a physical interpretation of this term is therefore beside the point. This point of view is awkward and more so if we look at Eq. (5) and assume that we have accelerated an orthogonal frame of reference. How do we justify covariant correction terms in an orthogonal spatial frame? Whichever way we twist this argument we either violate intuition or end up with terms that have no distinct physical meaning. These difficulties can be completely resolved by the covariant form (2b), where $\mathscr{G}$ represents the actual $\boldsymbol{D}$ and $\boldsymbol{H}$ field.

This problem of interpretation is intimately related to the following paradox. Consider a uniform magnetic induction $\boldsymbol{B}$ in freespace (no charge, no current) and a frame of reference spinning with angular velocity $\boldsymbol{\omega}$ so that $\boldsymbol{\omega} \| \boldsymbol{B}$. There is then a radial electric field $\boldsymbol{E} = \boldsymbol{v} \times \boldsymbol{B} = (\boldsymbol{\omega} \times \boldsymbol{r}) \times \boldsymbol{B}$ in the rotating frame.

The divergence of $\boldsymbol{E}$ in the rotating frame leads to a charge density

$$\varrho = \operatorname{div} \boldsymbol{E} = 2\boldsymbol{\omega} \cdot \boldsymbol{B}. \tag{6}$$

This conclusion is clearly absurd since it is at variance with our original assumptions. It results from a series of violations of the principle of general covariance. It is not true that the paradox is due to an inadmissible extrapolation of the Lorentz transformations for rotary motion. The existence of the electric field $\boldsymbol{E} = \boldsymbol{v} \times \boldsymbol{B}$ is a well-established fact in unipolar machines. Moreover, whether or not the Lorentz transformations apply is irrelevant from the point of view of general covariance because any transformation should preserve $\operatorname{div} \boldsymbol{D} = 0$. It will be clear that this paradox can easily be resolved, if we adhere to the covariant form (2b).

The first set of Maxwell equations

$$\operatorname{curl} \boldsymbol{E} = -\dot{\boldsymbol{B}}, \quad \operatorname{div} \boldsymbol{B} = 0, \tag{7}$$

is covariantly expressed by

$$\partial_{[\varkappa} F_{\lambda\nu]} = 0, \tag{8}$$

where F takes the place of $\boldsymbol{E}$ and $\boldsymbol{B}$. The brackets denote alternation over the indices. Eq. (8) is known to be invariant for general space-time transformations.

If we now assume that (2b) represents the physically correct form of the second set of Maxwell equations in general space-time coordinates, then we have in (8) and (2b) acquired a generally covariant set of equations independent of the metric. Any major physical conclusions derived by manipulating with (8) and (2b) will also have the covariant form, unless in the course of our derivation we violate the rules of covariance. An interesting case in point is the customary derivation of the energy-momentum law.

If we multiply (2b) by $F_{\sigma\lambda}$, sum over λ, and then use (8), we find

$$\partial_\nu (F_{\sigma\lambda} \mathcal{G}^{\lambda\nu}) + \tfrac{1}{2} \mathcal{G}^{\lambda\nu} \partial_\sigma F_{\lambda\nu} = F_{\sigma\lambda} c^\lambda. \tag{9}$$

The relation (9) is still a generally covariant relation, because in going from (2b) and (8) to (9) we have nowhere violated the rules of covariant operations.

The customary operation for obtaining from (9) the energy-momentum conservation relation then involves the somewhat questionable transition

$$\tfrac{1}{2} \mathcal{G}^{\lambda\nu} \partial_\sigma F_{\lambda\nu} = \tfrac{1}{4} \partial_\sigma (\mathcal{G}^{\lambda\nu} F_{\lambda\nu}) \tag{10}$$

and so the conservation relation becomes

$$\partial_\nu \{\tfrac{1}{4} \delta^\nu_\sigma \mathcal{G}^{\lambda\varkappa} F_{\lambda\varkappa} - F_{\lambda\sigma} \mathcal{G}^{\lambda\nu}\} = F_{\sigma\lambda} c^\lambda. \tag{11}$$

The lefthand side of Eq. (11) is not a vector density for general space-time transformations but the righthand side is. The root of this inconsistency lies in the transition (10), because (10) presupposes linearity and uniformity of the medium, that is, linearity and uniformity of the relation between $\mathcal{G}$ and F. Uniformity is expressed by the constancy of the coefficients relating $\mathcal{G}$ and F in a cartesian (uniform) frame of reference. The constancy of the coefficients relating $\mathcal{G}$ and F is clearly not a property that is preserved under arbitrary space-time transformations. Hence the transition (10) is not invariant under general space-time transformations.

It is possible to remedy the covariance defects of (10) and (11) by defining a Lagrangian density $\mathcal{L}$ such that

$$\mathcal{G}^{\lambda\nu} = \frac{\partial \mathcal{L}}{\partial F'_{\lambda\nu}}, \tag{12}$$

and writing the identity

$$\partial_\sigma \mathscr{L} = \partial_{(\sigma)} \mathscr{L} + \mathscr{G}^{\lambda\nu} \partial_\sigma F_{\lambda\nu}. \tag{13}$$

Substitution in (9) then yields the generally covariant conservation relation

$$\partial_\nu \{\delta^\nu_\sigma \mathscr{L} - F_{\lambda\sigma} \mathscr{G}^{\lambda\nu}\} = F_{\sigma\lambda} c^\lambda + \partial_{(\sigma)} \mathscr{L} \tag{14}$$

with two force terms on the righthand side. The first one represents the well-known Lorentz force, whereas the second one is associated with the radiation forces on medium inhomogeneities. The notation $\partial_{(\sigma)}$ designates a differentiation for constant F, hence a differentiation operating on the coefficients relating $\mathscr{G}$ and F.

We may refrain here from a more complete discussion of all the physical aspects of Eq. (14) [3]. The crucial point in the present context is that the insistence on general covariance brings out an essential physical feature because the radiation forces on medium inhomogeneities do cause an exchange between electromagnetic and mechanical energy and momentum. This exchange is expressed in (14) but missing in (11).

In the literature of physics it is customary to make (11) covariant simply by replacing the ordinary derivative by a covariant derivative. From the foregoing discussion it will be clear that there is no guarantee whatever that such a formal act is indeed meaningful from a physical point of view. It appears that this procedure leads to the same results as (14) for matterfree space, only if the relation between $\mathscr{G}$ and F is assumed to be linear; in a material medium it has no physical meaning.

4. Curvilinear Coordinates

It is traditional to identify general covariance with general space-time transformations, but one might just as well consider general covariance in a purely spatial sense, which means that the time does not occur in the admissible transformations. A description of the motion of a particle in the sense of the Hamilton-Jacobi theory is an example of a spatial, generally covariant, treatment. Developments of this sort really started with LAGRANGE and thus it might be said that the principle of general covariance for spatial coordinates originated some hundred years before EINSTEIN enunciated the principle of general covariance. The Hamilton-Jacobi method in mechanics really satisfied all requirements of spatial general covariance right from the start.

Parallel with the development of point mechanics there were also similar developments in the field theories of electromagnetics and continuum mechanics. The nature of certain problems in these field theories sometimes requires the use of curvilinear coordinates. For instance, when boundary conditions are defined on the surface of a sphere, it is advantageous to use spherical coordinates. The introduction of curvilinear coordinates in Maxwell theory and continuum mechanics has always been done in a somewhat ad hoc manner. There never was a coherent covariant treatment in the sense of the Hamilton-Jacobi theory. The whole problem centers on the question of how to transform expressions like the Laplacian, curl, and divergence.

A comparison between the treatment of general coordinates in the Hamilton-Jacobi theory and in field theory shows a rather striking difference in the manner of transforming the dependent variables. The invariant treatment of Hamilton-Jacobi leads quite naturally to the introduction of different vector species as dependent variables. The velocity vector in general coordinates is clearly a contravariant vector whereas the momentum vector appears as a covariant vector.

Traditionally, curl and divergence expressions are converted into curvilinear coordinates by using vector species of only one type as dependent variables. To avoid having to distinguish between vectors of different species, it is necessary to introduce local systems of orthogonal unit vectors. This is equivalent to introducing anholonomic coordinates, also called pseudo- or quasi-coordinates.

To illustrate the meaning of this statement, let us take the case of a transformation to spherical coordinates. The line element in spherical coordinates is given by

$$ds^2 = dr^2 + r^2 \sin^2\theta \, d\varphi^2 + r^2 \, d\theta^2. \tag{15}$$

Instead of the exact differentials dr, $d\varphi$, $d\theta$, it is now customary to introduce the inexact differentials

$$đ\xi = dr, \quad đ\eta = r \sin\theta \, d\varphi, \quad đ\zeta = r \, d\theta, \tag{16}$$

so that the line element expressed in these inexact differentials acquires the form of a line element with respect to an orthogonal cartesian frame. Thus,

$$ds^2 = đ\xi^2 + đ\eta^2 + đ\zeta^2, \tag{17}$$

and so no distinction is needed between different vector and tensor species (co- or contravariant, densities). Hence, the metric tensor associated with (17) is

$$g_{\lambda\nu} = \begin{pmatrix} 1 & 0 & 0 \\ 0 & 1 & 0 \\ 0 & 0 & 1 \end{pmatrix}. \tag{18}$$

Because of the inexactness of the differentials there are no integral variables ξ, η, ζ, and they are therefore called quasi-variables, pseudo-variables, or anholonomic coordinates.

As early as 1895, RICCI originated a covariant treatment of differential forms in anholonomic systems[1]. The coefficients of linear displacement are of course not given by the customary Christoffel symbols [as a matter of fact the latter vanish because of the constancy of (18)]; instead, they are given by the so-called Ricci coefficients of rotation.

Anholonomic coordinates are of a hybrid nature. It is not of course possible to express a dependent variable in the pseudo variables ξ, η, ζ. Holonomic variables must be resorted to, which in the case of the underlying system of spherical coordinates are conveniently taken as r, φ, θ. In an anholonomic system, references for the dependent and independent variables are different. This state of affairs issues from the voluntary restriction imposed on the choice of dependent variables, namely, the insistence on using only one vector species.

It seems appropriate to ask whether or not a freer choice of dependent variables would obviate the necessity of using anholonomic references. The answer is yes. There is no reason to retain the local frames of orthogonal unit vectors, if we stipulate that the curl operates only on covariant vectors and the divergence operates only on contravariant vector densities. If we then use the natural vector basis of the holonomic coordinate system, the curl and divergence expressions become so-called natural invariants independent of the metric tensor, and the irony is that neither the Christoffel symbols nor the Ricci coefficients are needed.

5. Intrinsic Transformation Behaviour and Physical Dimensions

Interesting questions arise when we look at the physical dimensional properties of the independent and dependent variables. In a Lorentz frame with cartesian spatial coordinates we associate a time dimension $[t]$ with the time coordinate, and a length dimension $[l]$ with the spatial coordinates. For a spatial transformation to spherical coordinates it is only the radial coordinate that has a length dimension; the two angle variables φ and θ are dimensionless. Hence, the dependent variables of a form-invariant expression must accordingly change their dimension so as to maintain the physical dimensional compatibility of the expression. Let us, for instance, compare the components of the curl of a vector in cartesian coordinates with the curl of the corresponding covariant vector in holonomic spherical coordinates:

$$\mathrm{curl}_y \boldsymbol{A} = \frac{\partial}{\partial z} A_x - \frac{\partial}{\partial x} A_z; \tag{19}$$

$$\mathrm{curl}_\varphi \boldsymbol{A} = \frac{\partial}{\partial \theta} A_r - \frac{\partial}{\partial r} A_\theta. \tag{20}$$

[1] See Reference [4], Chap. III, Sect. 9.

The expressions (19) and (20) clearly shown form invariance. For the cartesian expression (19) we know that the physical dimensions of A_x and A_z are the same. Thus,

$$[A_x] = [A_z]. \tag{21}$$

From (20), however, it follows that

$$[A_r] = [l^{-1}][A_\theta]. \tag{22}$$

The dimensional relation (22) is a typical consequence of the form congruence of (19) with (20). The result is quite common in Hamilton-Jacobi theory. If A_z is a linear radial momentum, then A_φ must have the dimension of an angular momentum as expressed by (22). Hence, the physical dimensional properties of vector components are not invariant under general coordinate transformations.

On the other hand, the customary treatment of curvilinear coordinates in field theory by means of anholonomic local frames of reference maintains the dimensional invariance of the vector components at the expense of the form invariance of the expression. The anholonomic curl expression corresponding to the holonomic expression (20) is known to be

$$\operatorname{curl}_\varphi \boldsymbol{A}' = \frac{1}{r}\left\{\frac{\partial A_r'}{\partial \theta} - \frac{\partial (r A_\theta')}{\partial r}\right\}, \tag{20a}$$

where

$$[A_r'] = [A_\theta'] \tag{23}$$

in accordance with (21). A further inspection of (20) and (20a) shows, as expected from (16), that the relation between the holonomic and anholonomic components is given by

$$\begin{aligned} A_r' &= A_r, \\ r A_\theta' &= A_\theta, \end{aligned} \tag{24}$$

where

$$r \operatorname{curl}_\varphi \boldsymbol{A}' = \operatorname{curl}_\varphi \boldsymbol{A}. \tag{24a}$$

Our next question is: Is it more important to maintain the form invariance of the equations or to maintain the physical dimensional invariance and homogeneity of vector components? We should remember that dimensional differences between vector components occur quite naturally, for instance in a properly covariant theory such as Hamilton-Jacobi mechanics. The dimensional difference between a component of angular momentum and one of linear momentum serves as an example. It would be clumsy to insist on clinging to the use of linear momentum when we treat a problem in polar coordinates because the polar coordinates are usually dictated by a rotational symmetry of the problem. Furthermore, a dimensional distinction between vector components becomes unavoidable in a space-time description. Consider the conjugate pairs of variables energy-time and momentum-space coordinates: a dimensional identification of length with time leads to an undesirable identification of energy with momentum.

We therefore take the stand that it is more important to maintain the form invariance of the equations and thus maintain and emphasize the dimensional individuality of dependent and independent variables as a means of identifying the correct physical nature of these variables.

It was mentioned earlier that form invariance requires a decision on the nature of the induced transformation behavior of the dependent variables. The transformation behavior that preserves the form (and the physical information associated with the form) will be referred to as the intrinsic transformation of the dependent variables. We have seen how form invariance is also associated with certain dimensional distinctions between the dependent variables. Hence the question: Is it possible to link the intrinsic transformation with the dimensional properties of dependent variables? The answer to this question is yes.

Briefly, this link can be established as follows. First we need a system of units that lends itself to a properly invariant space-time treatment. For this purpose it is useful to modify slightly the customary units of the mks system ($m=$ mass, $q=$ charge, $l=$ length, $t=$ time). Only the charge is an absolute invariant in the mks system; the mass, length, and time, each change when subjected to space-time transformations. A more desirable situation would be to have only the length and time changing, while the other two units remain unchanged. This situation can be obtained in principle, if we replace the mass unit m by an action unit $\hbar$. Since the action, according to its usual definition, depends of course on mass, length, and time; $(\hbar, q, l, t)$ is a properly independent set of units equivalent to the customary (m, q, l, t).

The $(\hbar, q, l, t)$ set offers the advantage that $\hbar$ and q are legitimate invariants under general space-time transformations but length and time are not. The invariance of $\hbar$ and q is suggested by the fact that they are in principle countable in terms of basic quantized units. The action and charge can be independent of length, time, and mass units if we simply state the number of elementary quanta of action and of charge.

The customary way of measuring length and time is a process of comparison with the divisions on a frame of reference. These divisions are generated by the natural basis of unit vectors. This natural basis as a rule changes from one space-time point to another and from one frame of reference to another. The preference is naturally for frames of reference with constant-base vectors, that is, provided such frames are allowed by the structure of the manifold (vanishing Riemann-Christoffel tensor).

The existing body of knowledge in physics strongly suggests that locally (that is, in the neighborhood of the space-time domain of our planet and era) we can obtain highly precise reproducible measurements of length and time that seem to have validity and meaning in a reasonably large space-time domain. Nevertheless, we know that these measurements are affected by the nature of a space-time structure. For instance, we know that an atomic or nuclear resonance frequency is only well defined if we extrapolate a local observation to a space-time point with a vanishing gravitational field. The recent checks of the gravitational red shift by means of the Mössbauer effect vividly illustrate the necessity for such extrapolations. The greater the accuracy of measurement the more it stresses local validity of results. The step from a subjective to an objective length or time measurement can be taken if we know how the measurement can be reduced to a space-time domain where a Lorentz-Minkowski structure prevails. On the other hand, the countable units $\hbar$ and q are unambiguous in meaning as invariants in a finite domain of the space-time manifold and are independent of its structural specification.

We now return to the relation between the induced (intrinsic) transformation behavior of dependent variables and their physical dimensions to discuss in particular the cases where the space-time form invariance can be preserved with dependent variables that belong to the species of tensors and vectors. Assuming for simplicity that all spatial coordinates have the same dimension, that of length, we can state the following three-part criterion for "good" physical tensor fields[1]:

a) All components have in common a factor (a) that is a rational and homogeneous algebraic function of $\hbar$ and q.

b) If the tensor field has a density of weight k, then all components also have a factor (b) of dimension $[l^3 t]^{-k}$.

c) The physical dimensions of the components themselves are obtained by multiplying the product of the two previous factors (a) (b) with an l or a t for every contravariant valence and with an l^{-1} or a t^{-1} for every covariant valence, depending on whether the index is a spatial or a time index.

Although it appears that a physically meaningful form invariance is always associated with good fields, it is course not true that good fields will automatically lead to form-invariant equations. The Maxwell-Minkowski fields $F_{\lambda\nu}$ and $\mathcal{G}^{\lambda\nu}$ meet the requirement that their transformations preserve the form of the field Eqs. (2b) and (8):

$$\begin{aligned}
F_{10} &\to \boldsymbol{E} \to [\hbar q^{-1}][l^{-1}t^{-1}] = [m l t^{-2} q^{-1}]\\
F_{23} &\to \boldsymbol{B} \to [\hbar q^{-1}][l^{-2}] = [m t^{-1} q^{-1}]\\
\mathcal{G}_{10} &\to \boldsymbol{D} \to [q][l^{-3}t^{-1}][lt] = [q l^{-2}]\\
\mathcal{G}^{23} &\to \boldsymbol{H} \to [q][l^{-3}t^{-1}][l^2] = [q l^{-1} t^{-1}].
\end{aligned}$$

[1] A three-dimensional version of this rule, based on spatially invariant units, is given by J. A. Schouten, Ref. [5], Chap. VI. A discussion of the space-time version based on the invariant units $\hbar$, q is given in E. J. Post, Ref. [7], Chap. II.

The operational importance of this rule of course hinges on the possibility of extracting a rational factor that depends solely on the invariant units $\hbar$ and q. We thus obtain the corollary.

Corollary. A tensorlike combination of dependent variables that does not allow the extraction of a rational homogeneous factor of $\hbar$ and q is not a "good" physical tensor field under general space-time transformations.

The following is a list of well-known classical fields that meet the criterion for a good field:

		[factor (a)]
$F_{\lambda\nu}$	***E***, ***B*** field	$\hbar/q$
$\mathcal{G}^{\lambda\nu}$	***D***, ***H*** field (density $k=1$)	q
$\mathcal{c}^{\lambda}$	charge current (density $k=1$)	q
A_{λ}	four-potential	$\hbar/q$
p_{λ}	energy-momentum vector	$\hbar$
$\mathcal{T}_{\nu}^{\lambda}$	energy-momentum tensor (density $k=1$)	$\hbar$
$\mathcal{f}_{\nu}$	power, force (density $k=1$)	$\hbar$

It is significant to note that the so-called power-force vector is not compatible with the requirements of the rule, but the power-force density vector is a "good" vector. Note also that the energy-momentum tensor is a mixed tensor density. The angular momentum conservation law cannot be expressed in terms of the symmetry of this tensor, because symmetry is not an invariant property of a mixed tensor. A tensor to which invariant symmetry conditions apply can be obtained by raising or lowering one of the indices by means of the metric tensor, and so an invariant expression for angular momentum conservation will necessarily include gravitational aspects.

It is important to note that it is not possible to extend the spatial angular momentum and torque vectors into a corresponding "good" four- or six-vector. The closest spacetime tensor construction that will encompass the classical concepts of torque and angular momentum is the so-called spin-tensor density

$$\mathcal{S}_{\varkappa}^{\lambda\nu} = -\mathcal{S}_{\varkappa}^{\nu\lambda}. \tag{25}$$

This tensor, which occurs in the modern literature, is used for symmetrization purposes of the energy-momentum tensor, because it has a vanishing double divergence, which follows from the skew symmetry expressed by Eq. (25). The dimensional structure of this tensor breaks up in the following manner:

$\mathcal{S}_{\varkappa}^{\lambda\nu}$	01 02 03	23 31 12
0	$[\hbar t^{-1} l^{-2}]$	$[\hbar t^{-1} l^{-2}]\,[l t^{-1}]$
1 2 3	$[\hbar l^{-3}]$	$[\hbar l^{-3}]\,[l t^{-1}]$

The three spatial rows (1, 2, 3) are readily interpreted as an angular momentum flow density. The zero row does not seem to be so open to easy interpretation, except that its divergence leads to an energy flow that is always conservative. The divergence of the spatial part correspondingly expresses a torque balance of the field.

The reader may have noticed that the possible existence of a metric field has not been emphasized in this discussion. The metric tensor is frequently used (in many cases, indiscriminately) for the so-called raising and lowering of indices of tensor fields. Since the metric tensor itself also represents a physical field, this raising and lowering of indices really amounts to a physical modification of the resultant field. The only situation in which the raising and lowering of indices does not affect the field physically is when the Riemann-Christoffel tensor of the metric vanishes. The metric tensor then contains solely structural information of the coordinate system with respect to the underlying Euclidean manifold. It is only then that the raising and lowering of indices becomes a purely geometric procedure, which enables us to map tensor species of the same valence onto each other. This procedure, however, muddles the distinction between the arbitrary coordinate information and the essential physical information in which we are interested. Hence — whether or not the Riemann-Christoffel tensor vanishes — it is of paramount importance to identify the physical field by its intrinsic transformation behavior.

Recognition of the metric as a physical field implies that the metric is also subject to the rule that links transformation and physical dimensions. The first question to be answered is, what is the invariant factor (a)? The previous discussion rules out the possibility of a line element in four-space with a length or time dimension; there is also no reason to consider an explicit $\hbar$ or q dependence. The only logically consistent possibility is a dimensionless line element in four-space. Hence, the factor (a) is dimensionless and the metric tensor decomposes dimensionally according to

$$g_{\lambda\nu} = \left(\begin{array}{c|c} t^{-2} & l^{-1}\,t^{-1} \\ \hline l^{-1}\,t^{-1} & l^{-2} \end{array}\right).$$

Just to touch upon an important class of dependent variables in modern physics — the spinors — can the induced transformation of spinors also be related to the physical dimensions of the fields they represent? The answer seems to be no; spinors are unambiguously related to the orthogonal space-time group. Thus, a discussion of spinors with reference to a general space-time frame necessarily involves the use of local orthogonal frames that are anholonomically related to the underlying general space-time frame. Spinors are analytically expressed in the variables of an orthogonal space-time frame and their components refer to a frame in the associated spinor space. If the local orthogonal space-time frames are essentially anholonomic because of a non-vanishing Riemann-Christoffel tensor, integral coordinate variables cannot be associated with them[1]. Hence, spinors cannot be analytically expressed in the holonomic coordinates of a curved space-time. This state of affairs is a serious operational and also epistemological limitation on the usefulness of the existing spinor concept for general relativity.

6. The Faraday-Schouten Geometric Images for Electromagnetic Fields in Space

SCHOUTEN showed that a "spatial reduction" of the so-called intrinsic transformation behavior of the "good" fields quite naturally leads to the same geometric illustrations of the electromagnetic field that were used by FARADAY. On this basis he extended FARADAY's work by further relevant detail. Prior to the era of relativity and the Ricci calculus, MAXWELL [7] had already added much relevant mathematics to FARADAY's concept. FARADAY had introduced the so-called tubes of flux to describe magnetic induction. Briefly, the greater the constriction of the tubes the higher the flux density. The intensity of the magnetic-induction vector is inversely proportional to the cross section of the tube. The geometric characteristics of the vector species are illustrated in Fig. 1. The arrow represents the circulation of the current generating the field, its exterior position denotes that the field is not affected by an inversion of coordinates[2].

---

[1] The absence of integral coordinate variables implies a situation in which we are dealing with so-called quasi-or pseudo-coordinates.

[2] These images should be considered as mnemonic devices illustrating the geometric properties of the fields in a heuristic sense. An extensive discussion can be found in Chapters II, III, VI, and VII of Ref. [5].

An example of a flow-density vector that changes sign when subjected to an inversion of coordinates is the electric current density. Its geometric image, given in Fig. 2, is similar to that of Fig. 1. The different behavior with respect to inversions is indicated by the interior position of the arrow.

The addition property of current density and time change of electric displacement shows that the electric-displacement vector should be

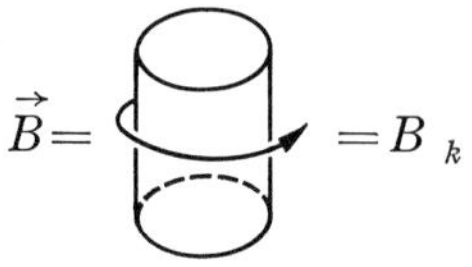

$\vec{B} =$ $= B_k$ $\vec{D} =$ $= \mathfrak{D}^l$

Fig. 1 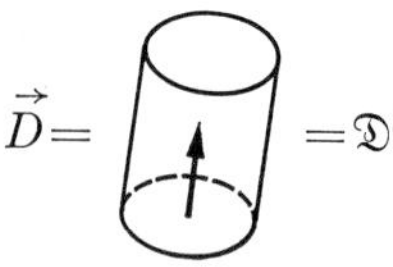 Fig. 2

Fig. 1. Geometric Illustration of the Magnetic-Induction Vector

Fig. 2. Geometric Illustration of the Electric Current-Density Vector and the Vector Density of Electric Displacement

represented by the same geometric image; the time derivative does not, of course, affect the spatial image.

The geometric image associated with the electric-field vector can be obtained from the equipotential surfaces surrounding a distribution of electrical charges. The electric field as a gradient of the potential is strong where the equipotential surfaces are close to each other. The

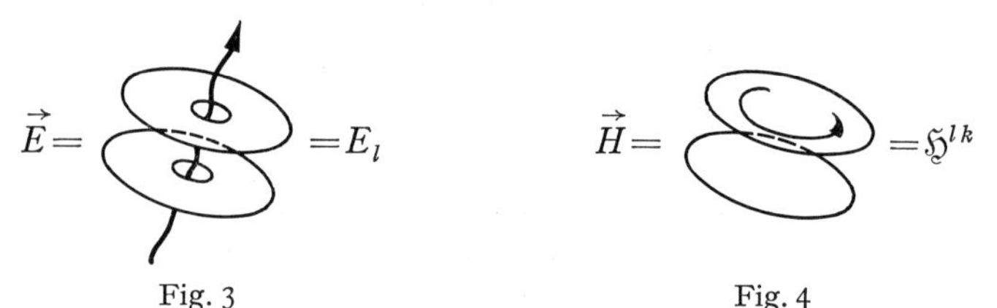

Fig. 3 Fig. 4

Fig. 3. Geometric Illustration of the Electric Field

Fig. 4. Geometric Illustration of the Magnetic Field

intensity is inversely proportional to the distance between adjacent potential surfaces. The arrow in Fig. 3 points from the high to the low potential.

The geometric image of the magnetic field vector is somewhat less obvious because the magnetic equivalent of electric charge does not exist. The image is almost the same as that of the electric field except that the arrow (Fig. 4) now denotes a circulation to account for the fact that currents, rather than charges, should be considered as the sources of the field. The magnetic field, like magnetic inductions, is accordingly not affected by a coordinate inversion.

SCHOUTEN noted that the vector species associated with his extension of the Faraday images shown in Figs. 1 to 4 are precisely those that lead to the spatial natural invariance of the Maxwell equations. SCHOUTEN'S observation is interesting because it shows that the experimentalist FARADAY had a greater subtlety of mathematical thought and insight than is apparent in contemporary mathematical treatment of electromagnetic theory.

It is now an easy matter to show that the spatial vector species (Figs. 1 to 4) are precisely those obtained from the "good" space-time fields.

The reduction is achieved simply by "freezing" the time index

$$F_{\lambda\nu} \begin{cases} \nearrow F_{0l} \rightarrow E_l \\ \searrow F_{lk} \rightarrow B_{lk} \end{cases} \qquad \lambda, \nu = 0, 1, 2, 3$$

$$\mathfrak{G}^{\lambda\nu} \begin{cases} \nearrow \mathfrak{G}^{0l} \rightarrow D^l \\ \searrow \mathfrak{G}^{lk} \rightarrow \mathfrak{G}^{lk} \end{cases} \qquad l, k = 1, 2, 3.$$

If we extract the density factor $g^{\frac{1}{2}}$ from the field $\mathfrak{G}^{\lambda\nu}$ according to

$$\mathfrak{G}^{\lambda\nu} = g^{\frac{1}{2}} G^{\lambda\nu}$$

[see Eq. (3)], we obtain the wrong type of vector images from the reduction

$$\mathfrak{G}^{\lambda\nu} \begin{cases} \nearrow G^{0l} \rightarrow \nearrow \\ \searrow G^{lk} \rightarrow \circlearrowright \end{cases}$$

The vector images obtained from $G^{\lambda\nu}$ by freezing the time index are those of the electric dipole (arrow) and magnetic dipole (surface area enclosed by current loop) instead of those of electric and magnetic polarization (dipoles/unit volume).

We thus see that the Faraday-Schouten geometrization of the electromagnetic field and the space-time invariant dimensional decomposition of the fields (the criterion for "good" physical tensor fields) together constitute a closely interrelated conceptual structure that corroborates the uniqueness of the natural invariance of the Maxwell equations.

7. The Riddle of Electromagnetic Mass

A problem that would be conspicuous by its absence from an anthology on covariance is that of electromagnetic mass. An electromagnetic interpretation of mechanical inertia had been suggested by ABRAHAM and LORENTZ in prerelativistic days. It lost most of its adherents, however, after the theory of relativity had established the relation between mass and energy as slightly different from the one obtained by ABRAHAM and LORENTZ (their inertial mass equaled $\frac{4}{3}$ of the electromagnetic energy divided by c^2.

One of the fallacies bred by the difference in the two results is that the discrepancy "clearly demonstrates" that at most only a part of the mass can be of electromagnetic nature. To bolster this conclusion, proponents submit that no part of the mass of neutral particles can be interpreted to be of electromagnetic origin. Had the Abraham-Lorentz derivation then been carefully checked for its relativistic validity, these arguments might have led to rather forbidding evidence against the thesis of electromagnetic mass. Forty years elapsed, however, after the publication of MINKOWSKI's papers on the invariant formulation of electrodynamics, before a flaw of invariance in the original Abraham-Lorentz treatment was explicitly corrected by KWAL [8]. The corrected calculation lead to the same relation between mass and energy as stated in the theory of relativity ($E=mc^2$)[1].

The mathematical differences between the two treatments are clearly illustrated by the relation

$$\begin{aligned} p_\nu &= \int \mathcal{T}_\nu^\lambda \, d f_\lambda \\ &= \underbrace{\int \mathcal{T}_\nu^0 \, d f_0}_{\text{Abraham-Lorentz term}} + \underbrace{\int \mathcal{T}_\nu^l \, d f_l}_{\text{terms added by KWAL}}, \end{aligned}$$

where

$\lambda, \nu = 0, 1, 2, 3$

$l = 1, 2, 3$

$0 =$ time index

$\mathcal{T}_\nu^\lambda =$ energy-momentum tensor

$p_\nu =$ energy-momentum vector

$d f_\lambda =$ element of three-dimensional hypersurface in space-time; hence, $d f_0$ is the purely spatial element.

The invariant treatment completely nullifies the original arguments against the electromagnetic interpretation of inertial mass. Every electromagnetic field that can be associated with a rest frame has its equivalent rest mass E/c^2. These includes electrostatic fields, magnetostatic fields, and standing waves, as well as suitable combinations of these that

[1] KWAL's short note remained unnoticed until ROHRLICH some ten years later independently introduced the corrected calculation [9, 10]. A recent textbook discussion is given by JACKSON [11]. FERMI [12] corrected the factor $\frac{4}{3}$ by an invariant treatment in 1922 although he did not address himself explicitly to the flaw in the Abraham-Lorentz treatment. An interesting but less convincing argument based on the indeterminateness of the Poynting vector was given by WILSON in 1936 [13].

permit a rest frame in which the radiation and corresponding recoil momentum vanish. A self-trapped standing wave would even allow an optional electromagnetic interpretation of the inertial mass of neutral particles.

These considerations of electromagnetic mass do, however, have a serious limitation from the point of view of general covariance. A domain vector of energy momentum as defined by KWAL's invariant integral relation loses its meaning in the context of general space-time transformations. KWAL's relation is only invariant within the framework of the full linear group, which of course covers the Lorentz group. It follows that any physical phenomenon essentially associated with the set of general space-time transformations, such as gravity, should be excluded from a discussion that is based only on linear invariance.

A physically meaningful general-covariant version of the invariant integral relation is no obvious and easy matter. Any electromagnetic theory of mass that describes inertial but not gravitational aspects of mass is still far from complete.

How a physically meaningful (nontrivial) general-covariant treatment might affect the riddle of electromagnetic mass is unknown — it has apparently never been attempted. A most curious fact is that it took over forty years to have corrected what seems to have been a simple relativistic flaw in the Abraham-Lorentz derivation.

8. Discussion

In his preface to a selection of papers on the subject of quantum electrodynamics, Schwinger [*14*] mentions the need for a version of the theory that manifests covariance at every stage of the calculations. This criticism is evidently directed against the standard practice of obtaining relativistic invariant theories by simply transcribing the end results derived from a spatial-invariant version — a method that may at times be adequate. In general, however, one should expect that rendering a theory in a relativistic invariant form is more than just transcribing the end results of a Newtonian version. The procedure of rewritting end results is at best no more than a tardy soothing of the physical conscience that reminds us physical laws should comply with relativity in some way or another.

SCHWINGER was referring to the requirement for Lorentz covariance. The remark is likewise applicable to situations where we expect the principle of general covariance to apply, and more so, because the chances of getting out of touch with physical reality are much greater here, especially if we proceed along the lines of the simple transcription method.

A more disciplined course than is now customary is required in the development of general-covariant methods. The actions recommended in this paper for achieving this goal neither violate nor restrict the application of well-known and accepted physical principles. On the contrary, they enhance the physical meaning, as is borne out by the improved epistemology of the physical fields. In distinguishing the disciplined from the undisciplined methods in general covariance, Schouten coined the term "method of natural invariance", which very appropriately expresses that disciplining indeed generates a more natural physical description.

The principle of general covariance became a principle of physical importance at the inception of general relativity. The further development of general relativity has been oriented in the sense of a geometric-topologic monism, and this has not produced any striking new physical aspects. Efforts to render the fundamental relations of quantum mechanics into a general-covariant form have also been unsatisfactory.

There is among physical theorists a great awareness of these existing limitations of invariance. The consensus is that the search should be for different types of invariances that match the typically different features of observability in microphysics; these invariances are not necessarily covered by the set of general space-time transformations. It would be straining a good point to assert dogmatically that all invariances in human observation are reducible to a space-time context.

At least one important reason for the frustrations encountered in general-covariant formulations in the realm of macrophysics lies in the almost exclusive use of the so-called covariant transcription method. Used in an indiscriminate manner, this method is dangerous in the sense that it bypasses epistemological aspects that should be the very basis for the construction of a meaningful general-covariant formulation. The examples discussed in this paper (Sec. 3) clearly demonstrate that the covariant transcription method leads to difficulties at very elementary levels. That elementary problems should be resolved before attacking sophisticated problems goes without saying. We cannot expect that new geometries and new topologies will reveal deeper insights into the physical world if we persist in ignoring the more elementary and relevant down-to-earth problems.

The recommendations (see Sec. 2) for promoting an improved implementation of general-covariant methods are by no means new. Neither are they sensational. They are simply guidelines for creating some order where at present no order exists. They do not confine physical imagination although they may restrict the use of some beloved toys that have by now proved to be without promise for the future.

ACKNOWLEDGMENTS

I am indebted to FRANCIS J. ZUCKER, Chief of the Waves and Circuits Branch of the AFCRL Microwave Physics Laboratory, for suggesting publication of a paper elaborating on the philosophy and unique value of the natural invariant method in electromagnetic theory. The discussions with those that helped to pinpoint communication weaknesses in earlier expositions of this subject matter are deeply appreciated.

I am indebted to B. S. KARASIK for editorial advice.

REFERENCES

[1] KRETSCHMANN, E.: Ann. Physik **53**, 575 (1917).
[2] EINSTEIN, A.: Ann. Physik **55**, 241 (1918).
[3] POST, E. J.: Phys. Rev. **133**, A 686 (1964).
[4] SCHOUTEN, J. A.: Ricci calculus, 2nd ed. Berlin-Göttingen-Heidelberg: Springer 1954.
[5] — Tensor analysis for physicists. Oxford 1949.
[6] POST, E. J.: Formal structure of electromagnetics. Amsterdam: North-Holland 1962.
[7] MAXWELL, J. C.: Proc. London Math. Soc. **3**, 224 (1871).
[8] KWAL, B.: J. phys. radium **10**, 103 (1949).
[9] ROHRLICH, F.: Am. J. Phys. **28**, 639 (1960).
[10] — Boulder lectures in theoretical physics, vol. II. New York: Interscience 1960.
[11] JACKSON, J. D.: Classical electrodynamics. New York: John Wiley & Sons 1962.
[12] FERMI, E.: Physik Z. **23**, 390 (1922).
[13] WILSON, W.: Proc. Phys. Soc. (London) **48**, 736 (1936).
[14] SCHWINGER, J.: Quantum electrodynamics. New York: Dover 1958.

Chapter 8

Foundation Problems in General Relativity[1]

Peter Havas

Department of Physics, Temple University
Philadelphia, Pennsylvania

Historically, EINSTEIN's theory of general relativity arose from attempts to fit a theory of gravitation into the framework of the space-time concepts of the theory of special relativity [*1*]. The special theory, in turn, arose from attempts to fit a theory of light into the framework of the space-time concepts of Newtonian physics. Within the latter, the existence of an absolute time and of an absolute meaning of simultaneity of distant events had been taken for granted; furthermore, MAXWELL's electromagnetic theory of light (or indeed any wave theory of light) seemed to associate a material medium with NEWTON's absolute space, and thus, in contrast to Newtonian point mechanics, to provide a physically preferred frame of reference. This appeared to offer the possibility of detecting effects due to motion with respect to the preferred frame. However, none of the expected effects could be observed, and it proved to be impossible to find an all-encompassing explanation for these negative results without transcending the bounds of the Newtonian concepts [*2*].

The fundamental physical fact which requires a reevaluation of these concepts is the existence of a maximum signal velocity in nature, the velocity of light in empty space. This was recognized by EINSTEIN in 1905, who also realized that the existence of this maximum velocity implies that the concept of simultaneity of distant events is not absolute, but involves an element of definition; then the most convenient way of establishing what is meant by "the same time" at widely separated points is by means of the fastest signals available. The adoption of the same definition in any inertial system allows the development of the special theory of relativity on the basis of EINSTEIN's two postulates, the "principle of relativity" (S): If properly formulated, the laws of physics are of the same form in all inertial systems, and the "principle of the constancy of the velocity of light" (L): In all such systems, the velocity of light in empty space has the same value c.

[1] Work supported in part by the National Science Foundation.

In the above, inertial systems are defined as in Newtonian mechanics as systems in which NEWTON'S first law is valid; within each such system, Euclidean geometry is assumed to hold. The first postulate (S) is simply an extension of the Newtonian principle of relativity from mechanics to all of physics. The break with Newtonian physics is not only in the addition of the second postulate (L) asserting the constancy of the velocity of light, but first and foremost in the recognition of this velocity as the maximum signal velocity in any frame of reference. The Newtonian belief in a physically meaningful absolute time can be more properly expressed from our present point of view as a belief in the existence of signals of arbitrarily high, even infinite, velocity. From this and the assumed validity of Euclidean geometry it follows that the Newtonian space-time coordinates in different inertial systems are related by the proper Galilei transformations. These transformations constitute a subgroup of a 10-parameter group, the Galilei group [*3*]. If the Newtonian principle of relativity is to hold, the equations of Newtonian physics must be invariant under this subgroup. However, in many instances the equations are invariant under the full group.

Similarly, EINSTEIN'S two principles allow the determination of the group of transformations relating the space and time coordinates in different inertial systems, the proper Lorentz group; this again constitutes a subgroup of a 10-parameter group, the Lorentz group, which reduces to the Galilei group in the limit $c \to \infty$ [*3*]. A number of important conclusions on space-time measurements can be drawn from the proper Lorentz transformations, such as the apparent contraction of moving objects, time dilatation, and the relativistic addition theorem of velocities [*4*].

It is essential for our later discussion to keep in mind that what is usually called the special theory of relativity consists of two parts of different standing [*5*]. The first one consists of the kinematical consequences of the space-time structure described by the proper Lorentz transformations. These are consequences which are assumed to hold for all macroscopic measurements; for any such measurements the theory contains clear prescriptions (the same prescriptions for all inertial systems in terms of the similarly defined space-time coordinates, including the use of Cartesian coordinates of the same "handedness").

The second part consists of specific theories or laws such as mechanics of mass points or of continuous media, electrodynamics, thermodynamics, quantum mechanics, or the conservation laws, which are all understood to conform to EINSTEIN'S two postulates, but are not direct consequences of them. They are thus not really integral parts of the theory of relativity, but are developed as required by theoretical considerations and experimental discoveries, and discarded if they are disproved by experiments or found theoretically barren. EINSTEIN'S principle of relativity requires their invariance under the proper Lorentz group, but frequently invariance under the full group exists or is assumed. Some theories and laws like the mechanics of a single particle, macroscopic electrodynamics, conservation of energy-momentum and as its consequence the equivalence of energy and inertial mass, are very well established. Others, such as relativistic thermodynamics, have not yet been thoroughly investigated and understood; still others, such as quantum field theory, have had some successes, but have not been proved to be mathematically consistent. A striking example of a "law" once widely assumed to be generally valid which later had to be discarded is the "conservation of parity," which rested on invariance under *improper* Lorentz transformations, a requirement *not* imposed by the principle of relativity.

The requirements of relativity leave a wide latitude to the possible form and content of physical theories. Clearly they permit both classical and quantum theories. In spite of the widely held belief that they impose the necessity of theories of near action (field theories), it has been well established that they do *not* preclude action-at-a-distance formulations [*6*]. They *do* exclude Newtonian action-at-a-distance theories, however, such as NEWTON'S theory of gravitation. But it is not at all difficult to develop theories of gravitation which do conform to the relativistic requirements, and which reduce to NEWTON'S theory in the limit $c \to \infty$ [*1*, *7*].

The latter theory had been very successful in explaining problems in celestial mechanics, except for an anomaly in the motion of Mercury. After all known perturbations were taken into account, there remained an unexplained advance of the perihelion of 43''/century [*8*], and it was natural to hope that a new theory of gravitation would not only satisfy the formal requirement of Lorentz invariance, but would also succeed in solving the one problem for which Newtonian theory had failed.

It is fortunate for the development of physics that the early Lorentz-invariant theories of gravitation were not able to accomplish this [*1*], but only some of those devised after EINSTEIN had developed his general theory [*7*]; otherwise their success might have discouraged a study of EINSTEIN'S radically new ideas and the mathematical complexities of their realization. However, the general theory not only provided a satisfactory explanation of the anomaly, but it also predicted two further effects which subsequently were to some degree established experimentally, and thus offer a new challenge to Lorentz-invariant theories[1].

[1] Although some of these theories were indeed able to arrive at an explanation of all three effects, this does not arise in such a natural manner as in EINSTEIN'S theory, and is therefore widely considered to be less satisfactory. This point is discussed in some detail in a review article by A. SCHILD in [*9*], p. 69.

Within three years of the creation of the special theory of relativity, EINSTEIN had come to the conclusion that the development of a satisfactory theory of gravitation would require more than a change from Galilei covariance to Lorentz covariance [*10*]. This conclusion was rooted in several considerations, which were stressed with varying emphasis in the next seven years on the tortuous road to the final form of his general theory of relativity. It will be one of the main tasks of this talk to examine to what extent these considerations can still be considered as valid and compelling, to what extent they determine the form of the theory, and whether they can be replaced in whole or in part by other considerations which would still lead to the same form of the theory. We shall then have to examine other problems inherent in the foundations of the theory as formulated by EINSTEIN, and the extent to which the considerations and achievements are unique to his theory. Clearly time does not permit a full discussion of all these points, and frequently I will be able only to sketch an argument or to state a conclusion without any discussion.

Neither in Newtonian mechanics nor in special relativity is there a single preferred reference system. However, in both these theories there is a class of preferred systems, the inertial systems, related to each other by motions with constant velocity. Neither theory offers an explanation for the existence of this privileged class[1], and the very definition of this class by means of NEWTON'S first law is subject to difficulties, because the universal nature of gravitation does not permit the absence of external forces on a body, except if it is infinitely distant from all other bodies. ERNST MACH'S suggestion that inertial effects should be ascribed to the influence of distant masses rather than taken as properties of space ("MACH'S principle") was taken by EINSTEIN to mean that "the laws of physics should be such that they are valid in frames of reference in arbitrary motion" [*14*]. However, he realized (by considering a frame of reference in uniform rotation relative to an inertial system) that if this "principle of general relativity" was to hold and the special theory of relativity was to be valid in the absence of gravitational fields, it was not always possible to maintain the direct physical significance of space-time coordinates familiar from Newtonian and special relativistic theory, i.e. to define coordinates globally in such a manner that spatial

[1] The most extensive and best known prerelativistic criticism of the Newtonian concepts is due to ERNST MACH [*11*]. However, NEWTON'S ideas on absolute space were already criticized by GEORGE BERKELEY in the 18th century, and CHRISTIAAN HUYGENS, a contemporary of NEWTON, insisted that mechanics should be describable in terms of relative motion only; for details see [*12*]. A transcription of Newtonian mechanics into relative coordinates on the basis of a hypothesis of FÖPPL (whose work, like MACH'S, had considerable influence on EINSTEIN'S thoughts) was given in [*13*].

and temporal coordinate difference could be measured directly with standard measuring rods and clocks. Thus he abandowed this requirement and postulated the "principle of general covariance" (C): If properly formulated, the laws of physics are of the same form in all coordinate systems.

The foregoing considerations are actually independent of gravitation. However, another aspect of the consideration of relative motions of frames of reference other than with constant velocity provided EINSTEIN with the link to gravitation. In Newtonian theory one has to distinguish between inertial mass (which is involved in NEWTON'S second law, $\boldsymbol{F}=m\boldsymbol{a}$) and gravitational mass (which enters NEWTON'S law of gravitation). It is an experimental fact that the ratio of these two masses is the same for all bodies. This had been verified with an accuracy of 1 part in 10^8 by EÖTVÖS[1], who made use of the fact that "inertial forces," i.e. forces on a body due exclusively to the use of a noninertial frame of reference in Newtonian mechanics, are (by definition) proportional to the inertial mass of that body.

Because of the universal proportionality of the Newtonian inertial and gravitational masses it is not possible to distinguish by mechanical experiments between an inertial frame of reference with a homogeneous gravitational field in some direction, and a frame of reference which is accelerated in the opposite direction with respect to an inertial system without gravitational field (EINSTEIN'S elevator). EINSTEIN realized first of all that if the equivalence of these systems was *assumed*, it could be proved that the ratio of inertial and gravitational mass was the same for all bodies, even taking into account the result of special relativity that the inertial mass of a body was not necessarily constant. Furthermore he realized that (given the finite velocity of light) without this assumption it might be possible to distinguish the two reference system by optical experiment, while this assumption entailed the bending of light in a gravitational field and a dependence of the rate of a clock on the gravitational potential [*16*]. Thus he was led to postulate the "principle of equivalence" (E), which can be expressed thus: The local effects of a gravitational field are equivalent to those appearing in the description of physical phenomena relative to an accelerated frame of reference.

It is frequently asserted or implied that principles (C) and (E) by themselves lead to EINSTEIN'S theory of general relativity. This is not the case, however. First of all, it was pointed out as early as 1917 by

[1] The numerous papers by EÖTVÖS on gravitation are collected in [*15*]. It should be noted that the accuracy of EÖTVÖS' torsion balance experiments was not even approached by that of any others until the nuclear resonance experiments of the 1940's, and thus the experimental basis of general relativity was far more securely established than that of any contemporaneous theory.

KRETSCHMANN [*17*] and concurred in by EINSTEIN [*18*] that the principle of general covariance is devoid of physical content, and any theory whatever can be formulated in a generally covariant form; in his later writings EINSTEIN considered this principle to be mainly of heuristic value in helping select a theory which is of a simple form if written in generally covariant form [*10, 19*]. Both KRETSCHMANN and EINSTEIN thought that a generally covariant formulation of Newtonian mechanics would be extremely complicated; it was realized later, however, that this was not at all these case[1], and that NEWTON's theory of gravitation could be put into a form very similar to that of EINSTEIN [*22, 23*]. We shall return to this question later.

However, we note that the fact that the principle of general covariance is physically empty also disposes of EINSTEIN's original belief that its imposition would *automatically* incorporate MACH's principle into the theory [*14*]. Whether or not the theory in its final form actually contains MACH's principle is a separate question to be considered later.

We now turn to the principle of equivalence. It is a very important consequence of this principle — or rather a restatement of the principle in a different form — that locally (i.e. in the infinitesimal neighborhood of a point) a gravitational field can be "transformed away", i.e. new reference systems (accelerated with respect to the one originally considered and realizable by a freely falling system) can be introduced in which all effects depending on gravitational fields alone (but not effects depending on derivatives of the field!) are absent. It is these new systems which are called inertial systems in EINSTEIN'S theory of relativity; the possibility of giving a local definition of inertial system and a prescription for realization of such a system is generally considered to be a major achievement of this theory (removing the difficulties associated with the verification of a system being inertial by means of NEWTON's first law, due to the universal presence of gravitational forces).

It is rather surprising that it was only pointed out quite recently that the *same* redefinition of an inertial system is possible within the framework of Newtonian mechanics [*24, 23*]. It also seems to have been overlooked that the realization of such an inertial system by a system of coordinates attached to a freely falling body is not devoid of difficulties. There is of course no problem if the only forces acting on the falling body are gravitational. This could be assured by removing the

[1] Some aspects of a four-dimensional generally covariant development of point mechanics are contained in the use of homogeneous coordinates in Lagrangian dynamics (compare the article by J. L. SYNGE in [*20*], part E II); however, this is usually treated as a purely formal device. On the other hand, much work has been done on the generally covariant formulation of the classical mechanics of continua following the early work of CARTAN [*21*], and a comprehensive review is presented by C. TRUESDELL and R. TOUPIN in [*20*].

falling body sufficiently far from all other bodies — the same solution of our dilemma as in the case of the original Newtonian problem, but certainly a solution which requires more than local considerations and which would renounce the universal applicability of the prescription. If we wish to maintain this universality, we have to insist on free fall even in the possible presence of nongravitational forces (whose possible sources are *not* to be considered in a *local* experiment). This requires that the body to be dropped possess no nongravitational properties (be uncharged, nonmagnetic, etc.), which can only be established by dropping it and observing its motion. This is a circular argument of the same type as appears in the case of a local test of an inertial system by means of the first law in the original Newtonian theory.

Regardless of these practical difficulties the principle of equivalence assures us by definition of the existence of a class of inertial systems at any point in space-time. However, the connection of this choice of language with physics is assured only by imposing the requirement that these inertial systems possess some of the properties of inertial systems as defined previously; thus a minimum requirement is that in these systems NEWTON'S first law should hold, a requirement common to Newtonian physics and to the special theory of relativity[1]. EINSTEIN chose as a further requirement to demand the validity of the special theory of relativity (and in particular of its space-time structure). However, as we wish to explore to what extent the principles (C) and (E) determine a theory, we have to consider also the alternative of demanding the validity of Newtonian physics instead [*23*].

To proceed further we have to study the meaning of the principle (E) in more detail. Following its originator, we introduced it by a consideration of "EINSTEIN'S elevator" which showed that in Newtonian mechanics *gravitational* forces can be replaced by properly chosen *inertial* forces. It is frequently asserted that (E) only extends this equivalence to nonmechanical phenomena.

If this is all that (E) is understood to contain, and we recall that (C) is physically empty, then, unlike the relation between the fundamental principles of Newtonian mechanics and of *special* relativity, the basic principles of Newtonian mechanics and of *general* relativity are compatible.

On the other hand, it is also frequently asserted (or tacitly assumed) that *any inertial* force can be replaced by a suitably chosen *gravitational* force. This statement does not follow from EINSTEIN'S elevator or similar illustrations, and indeed is not true for the usual formulations of Newtonian mechanics. Nevertheless, another formulation of Newtonian

[1] This requirement can not be avoided if we wish to apply the theory to macroscopic phenomena. A weakening in the microscopic domain, unobservable macroscopically, can be considered and will be discussed later.

mechanics (identical in physical content with the usual one) can be given [23] which does conform to this interpretation of (E), as will be discussed later. Thus even with this more stringent interpretation of the principle of equivalence we shall find Newtonian mechanics to be compatible with the basic principles of general relativity.

Up to this point we could discuss all questions without explicit need for any complex mathematical apparatus, but we cannot proceed further without some technical equipment, specifically in differential geometry [25]. Principle (C) requires the consideration of arbitrary coordinate transformations

$$x'^{\mu} = x'^{\mu}(x^{\varrho}), \qquad \varrho, \mu = 0, 1 \ldots n-1, \tag{1}$$

in a space of n dimensions. In this space a contravariant vector A^{ϱ} is defined as a quantity with n component which transform like the $d x^{\varrho}$, i.e.

$$A'^{\mu} = \frac{\partial x'^{\mu}}{\partial x^{\varrho}} A^{\varrho}. \tag{2}$$

Thus the coordinate differentials $d x^{\varrho}$ form a contravariant vector, but for the finite coordinate differences Δx^{ϱ} this is only the case if the transformation (1) is linear, and for the coordinates x^{ϱ} themselves only if it is also homogeneous[1]. A quantity B_{ϱ} is called a covariant vector if its components transform as

$$B'_{\mu} = \frac{\partial x^{\varrho}}{\partial x'^{\mu}} B_{\varrho}. \tag{3}$$

In general one can *not* unambiguously associate a covariant and a contravariant vector with each other.

We can define covariant, contravariant, and mixed tensors of any rank by similar expressions, i.e.

$$T'^{\alpha\beta\ldots}{}_{\mu\nu\ldots} = \frac{\partial x'^{\alpha}}{\partial x^{\varkappa}} \frac{\partial x'^{\beta}}{\partial x^{\lambda}} \cdots \frac{\partial x^{\varrho}}{\partial x'^{\mu}} \frac{\partial x^{\sigma}}{\partial x'^{\nu}} \cdots T^{\varkappa\lambda\ldots}{}_{\varrho\sigma\ldots}; \tag{4}$$

a vector is a tensor of rank 1, a scalar (invariant) one of rank 0.

We can also define addition and subtraction for tensors of the same rank and multiplication and contraction for tensors of any rank. To define differentiation maintaining the tensor character of any expression, we have to introduce an *affine connection* $\Gamma^{\sigma}_{\varkappa\lambda}$ with the inhomogeneous transformation law

$$\Gamma'^{\varrho}_{\mu\nu} = \frac{\partial x'^{\varrho}}{\partial x^{\sigma}} \frac{\partial x^{\varkappa}}{\partial x'^{\mu}} \frac{\partial x^{\lambda}}{\partial x'^{\nu}} \Gamma^{\sigma}_{\varkappa\lambda} + \frac{\partial x'^{\varrho}}{\partial x^{\sigma}} \frac{\partial^2 x^{\sigma}}{\partial x'^{\mu} \partial x'^{\nu}}; \tag{5}$$

[1] This introduces a slight notational inconsistency which could be avoided only at great inconvenience. A similar (standard) inconsistency is the notation for the $\Gamma^{\sigma}_{\varkappa\lambda}$ introduced below, which transform according to Eq. (5).

thus $\Gamma^{\sigma}_{\varkappa\lambda}$ does not transform as a tensor except under *linear* transformations. Then we can define covariant derivatives by

$$T^{\varkappa\lambda\ldots}{}_{\varrho\sigma\ldots;\nu} = \partial_\nu T^{\varkappa\lambda\ldots}{}_{\varrho\sigma\ldots} + \Gamma^{\varkappa}_{\alpha\nu} T^{\alpha\lambda\ldots}{}_{\varrho\sigma\ldots} + \\ + \Gamma^{\lambda}_{\alpha\nu} T^{\varkappa\alpha\ldots}{}_{\varrho\sigma\ldots} + \cdots - \Gamma^{\alpha}_{\varrho\nu} T^{\varkappa\lambda\ldots}{}_{\alpha\sigma\ldots} - \Gamma^{\alpha}_{\sigma\nu} T^{\varkappa\lambda\ldots}{}_{\varrho\alpha\ldots} - \cdots, \\ \partial_\nu \equiv \frac{\partial}{\partial x^\nu}. \tag{6}$$

The affine connection can be arbitrarily assigned in one coordinate system; it determines the meaning of parallel displacement in the space considered, which is called an affine space. In the following we shall only be concerned with symmetric connections $\Gamma^{\sigma}_{\varkappa\lambda} = \Gamma^{\sigma}_{\lambda\varkappa}$. It is clear from Eq. (5) that if we have two different affine connections, their difference transforms like a tensor.

If we form the second covariant derivatives by successive applications of Eq. (6), then in general the differentiations with respect to different coordinates do not commute. This property of the affine space is called curvature and is described by the Riemann-Christoffel curvature tensor $R^{\varkappa}{}_{\mu\lambda\nu}$ and the contracted curvature tensor $R_{\mu\nu}$ defined as

$$R^{\varkappa}{}_{\mu\lambda\nu} \equiv \partial_\lambda \Gamma^{\varkappa}_{\mu\nu} - \partial_\nu \Gamma^{\varkappa}_{\mu\lambda} + \Gamma^{\varkappa}_{\varrho\lambda} \Gamma^{\varrho}_{\mu\nu} - \Gamma^{\varkappa}_{\varrho\nu} \Gamma^{\varrho}_{\mu\nu}, \qquad R_{\mu\nu} \equiv R^{\lambda}{}_{\mu\lambda\nu}. \tag{7}$$

If $R^{\varkappa}{}_{\mu\lambda\nu} = 0$, the space is called flat.

A geodesic in affine space is defined as the straightest line. As a function of a parameter τ which is defined up to a linear transformation it satisfies the equation

$$\frac{d^2 x^\varrho}{d\tau^2} + \Gamma^{\varrho}_{\mu\nu} \frac{dx^\mu}{d\tau} \frac{dx^\nu}{d\tau} = 0. \tag{8}$$

In general, length can only be defined along a geodesic, and lengths along different geodesics can not be compared. Such a comparison becomes possible in general only if one can indroduce a symmetric tensor of rank two $g_{\mu\nu}$ with vanishing covariant derivatives, the metric tensor. Then the distance is determined by

$$ds^2 = g_{\mu\nu} dx^\mu dx^\nu, \tag{9}$$

which by suitable choice of the arbitrary factor in τ agrees with the distance along the geodesics.

If the determinant of the $g_{\mu\nu}$ vanishes, the metric is called singular; if it does not vanish anywhere, the space is called metric or Riemannian. In such a space the geodesics are the shortest (or at least extremal) as well as the straightest lines; the equations $g_{\mu\nu;\varrho} = 0$ can be resolved for the affine connections, which are given by

$$\Gamma^{\varrho}_{\mu\nu} = \left\{ {\varrho \atop \mu\nu} \right\}, \\ \left\{ {\varrho \atop \mu\nu} \right\} = \tfrac{1}{2} g^{\varrho\sigma} (\partial_\mu g_{\nu\sigma} + \partial_\nu g_{\sigma\mu} - \partial_\sigma g_{\mu\nu}), \tag{10}$$

where $\left\{ {\varrho \atop \mu\nu} \right\}$ is called the Christoffel three-index symbol and $g^{\mu\nu}$ is the inverse of $g_{\mu\nu}$, defined by

$$g_{\mu\varrho} g^{\varrho\nu} = \delta_\mu^\nu, \tag{11}$$

where δ_μ^ν is the Kronecker δ (which transforms as a mixed tensor). With the help of these two tensors we now can associate with each contravariant vector A^ϱ a covariant vector $A_\mu = g_{\mu\varrho} A^\varrho$, and conversely $A^\varrho = g^{\varrho\mu} A_\mu$.

We do not really need all this machinery to discuss (or even to deal mathematically with) the special theory of relativity. However, as was first realized by the mathematician MINKOWSKI [26] it is very convenient to formulate this theory in a space of four dimensions, obtained by combining the time and the three space coordinates, with

$$x^0 = t, \qquad x^1 = x, \qquad x^2 = y, \qquad x^3 = z. \tag{12}$$

This space can be taken to be metric and flat; if only Cartesian spatial coordinates [and thus only *linear* transformations (1), equivalent to the proper Lorentz transformations] are considered, the affine connections corresponding to Euclidean geometry vanish in all permissible coordinate systems. The metric tensor $g_{\mu\nu}$ can be chosen to have the components

$$g_{00} = 1, \quad g_{11} = g_{22} = g_{33} = -c^{-2}, \quad g_{\mu\nu} = 0 \quad \text{for} \quad \mu \neq \nu. \tag{13}$$

As was realized by various authors [3, 21, 22], the space-time of Newtonian physics can also be considered as a space of four dimensions, with the same convention (12). This space is flat, like that of special relativity, and similarly, if only Cartesian spatial coordinates are considered, the affine connections vanish in all permissible coordinate systems. However, it is not possible to introduce a physically meaningful nonsingular metric; we can instead choose a metric tensor $g_{\mu\nu}$ with components

$$g_{00} = 1, \quad \text{all other} \quad g_{\mu\nu} = 0, \tag{14}$$

which has no inverse, and a second singular tensor $h^{\mu\nu}$ with components

$$h^{11} = h^{22} = h^{33} = -1, \quad \text{all other} \quad h^{\mu\nu} = 0, \tag{15}$$

which permits the definition of a nontrivial interval in the case $dx^0 = 0$ in which Eq. (9) with (14) gives zero regardless of the spatial separation. The space characterized by (14) and (15) is affine, but not metric. We note that

$$g_{\mu\varrho} h^{\varrho\nu} = 0. \tag{16}$$

In *both* spaces, the vanishing of the acceleration of a free particle with coordinates z^ϱ can be described by

$$\frac{d^2 z^\varrho}{d\tau^2} = 0, \tag{17}$$

with

$$d\tau^2 = g_{\mu\nu}\, dz^\mu\, dz^\nu. \tag{18}$$

Comparing this with Eq. (8), we see that in both cases the particle moves along a geodesic.

The formulation just given is not generally covariant, because only linear transformations are allowed. It can be made generally covariant, however, by *defining* $g_{\mu\nu}$ and $h^{\mu\nu}$ as tensors with vanishing covariant derivatives [which take on the special values (13) or (14) and (15), respectively, in those coordinate systems in which the affine connections vanish] for *arbitrary* transformations and replacing (17) by the general equation of a geodesic

$$\frac{d^2 z^\varrho}{d\tau^2} + \Gamma^\varrho_{\mu\nu} \frac{dz^\mu}{d\tau} \frac{dz^\nu}{d\tau} = 0 \tag{19}$$

maintaining (18). In analogy to Newtonian usage, the second term can be called an inertial acceleration, but it should be noted that it may appear simply because of the use of ordinary non-Cartesian coordinates rather than of frames of reference accelerated relative to an inertial frame. In deference to the principle of equivalence we can also call it a gravitational acceleration, but we must realize not only that this stretches the principle to go beyond the equivalence with accelerated frames of reference, but that there exists an entire class of frames of reference in which the gravitational field vanishes *everywhere*, the frames of reference considered originally. Expressed differently, the spaces previously considered were flat, and this property is not changed by allowing arbitrary coordinate transformations, the vanishing of the curvature tensor $R^\varkappa_{\mu\lambda\nu}$ being a tensor equation.

The giant step taken by EINSTEIN was to assume that the restriction to flat space could be dropped, and real gravitational fields (i.e. fields which can be transformed away *only locally*) could be described by a four-dimensional space with nonvanishing curvature tensor. All of the formalism just considered could be maintained, including the form (19) of the equation of motion of a particle in a gravitational field, the geodesic law. The space to be considered was the metric space characterized by the invariant distance (9), where at any point the metric tensor $g_{\mu\nu}$ could be chosen to take the values (13) and the Γ's could be chosen to be zero.

As was only realized later [*22, 23*], it is possible to arrive at a "generally relativistic" Newtonian theory of gravitation (identical in physical content with the usual theory) by maintaining all of EINSTEIN's requirements except the last one; instead, the space to be considered is the affine space characterized by the invariant distance (9) with the singular metric tensor $g_{\mu\nu}$ and a second singular tensor $h^{\mu\nu}$, where at any point these tensors can be chosen to take the values (14) and (15), and the Γ's can be chosen to be zero. However, because $g_{\mu\nu}$ is singular, the affine connections Γ now are not expressible in terms of the metric tensor, unlike the case of EINSTEIN'S theory. The formal connection between the two theories becomes more apparent if we introduce a tensor $H^{\mu\nu} = g^{\mu\nu}/c^2$ in EINSTEIN'S theory. Then Eq. (11) becomes

$$g_{\mu\varrho} H^{\varrho\nu} = c^{-2} \delta^{\nu}_{\mu} \tag{20}$$

which degenerates into Eq. (16) in the limit $c \to \infty$.

Thus, far from being a complete break with Newtonian theory, EINSTEIN'S assumption can be considered as a natural generalization of this theory, the essential physical difference being only the existence of a limiting signal velocity c, exactly as in the absence of gravitation.

EINSTEIN'S assumption establishes a connection between a physical quantity (the gravitational force) and geometric quantities (the affine connections and the curvature tensor determined by them). A number of important conclusions can be drawn from this association without the necessity of considering the relation between the gravitational field and its sources. The geodesic law allows a detailed discussion of the mechanics of a single body, and its extension to light (requiring it to follow a null geodesic) permits the discussion of optical effects. In the case of EINSTEIN'S theory this can be found in all textbooks [*27*]; the Newtonian case has also been discussed recently [*23*].

In both cases geometry and physics are interrelated, and in both cases the affine connections and the curvature tensor can be determined experimentally from a study of *gravitational* effects. But as in EINSTEIN'S theory the affine connections can be expressed in terms of the metric tensor, the curvature tensor can in principle also be determined from a study of *metric* effects; this is not possible in the Newtonian case [*23*]. Another way to look at this distinction is to recall that geometry in the ordinary sense refers to properties of the three-dimensional subspace $x^0 =$ constant of the four-dimensional space-time. This geometry may be non-Euclidean even if the 4-space is flat. A case in point is the space-time of special relativity, even though the spatial geometry within an inertial system is Euclidean by assumption. If we allow arbitrary coordinate transformations (including the time), the square of the three-dimensional line element involves more than just the spatial components of the metric tensor, and is not necessarily reducible to a sum of squares everywhere by a transformation of the spatial coordinates alone; thus the spatial geometry may be non-Euclidean, the geometry of a rotating disk being the most familiar example [*27*]. On the other hand, in the Newtonian case the geometry of the three-dimensional subspace is determined by the tensor $h^{\mu\nu}$, and it can be shown that there always exist global coordinate systems in which it takes the values (15), which correspond to Euclidean geometry [*23*].

Up to this point we have explored the space-time of the general theory of relativity as determined by EINSTEIN's principles (C) and (E) together with additional interpretations and assumptions. The main point in our discussion was that an essential step beyond the two principles had to be taken by assuming that a number of relations and concepts valid in a *flat* four-dimensional space could be maintained in a curved space, whose curvature corresponds to real gravitational fields. We now turn to the problem of establishing the equations relating the gravitational effects to their sources. Since from Eq. (19) these effects are described by the affine connections $\Gamma^{\varrho}_{\mu\nu}$, we have to find the equations obeyed by the Γ's.

EINSTEIN had associated a *metric* space with the affine connections, which thus from Eq. (10) are functions of the metric tensor $g_{\mu\nu}$. From a study of the nonrelativistic limit of the geodesic law (19) it is apparent that in this limit the component g_{00} is proportional to the Newtonian gravitational potential U. Thus, the successive attempts of EINSTEIN to establish the equations for the Γ's or equivalently for $g_{\mu\nu}$, were guided by the Newtonian Poisson equation for the gravitational potential

$$\nabla^2 U = 4\pi G \varrho, \tag{21}$$

which relates U to the mass density of matter ϱ and the gravitational constant G. Since the special theory of relativity had associated mass with all forms of energy (and the principle of equivalence implies the association of gravitational effects with all mass), this suggested a replacement of the mass density ϱ by the total energy density T_{00}. But if the equations were to be manifestly covariant under arbitrary coordinate transformations, they could not single out the 0-component of a quantity, and thus EINSTEIN assumed that the right-hand side of Eq. (21) should be replaced by an expression proportional to the (symmetric) total energy-momentum tensor $T_{\mu\nu}$ of matter as well as of all nongravitational fields. The apparent exclusion of the gravitational field itself as a source of gravitation will be discussed later; the inclusion of a matter tensor, on the other hand, which requires more for its description than field variables, was always considered by EINSTEIN to be a makeshift solution, to be ultimately replaced by a theory of matter in terms of fields alone[1].

---

[1] The search for a field theory of matter was the main motivation for EINSTEIN's development of his unified field theory, which is discussed in detail in [*28*] and [*29*]; however, in spite of many attempts, he was not able to establish the existence of nonsingular particle-like solutions. The possibility of obtaining particles in the general theory of relativity by the use of non-Euclidean topology was suggested by EINSTEIN and ROSEN [*30*]; for a summary of later developments along this line see [*31*]. Whether EINSTEIN's program of a pure field theory of matter can be realized remains an open question, however.

The right-hand side of the proposed equations being a symmetric tensor of rank two, the same tensor character must be required of the left-hand side. Furthermore, the left-hand side of Eq. (21) being a partial differential expression linear in the second derivatives of U, EINSTEIN required as the simplest possibility that the left-hand side of the new equations should be a partial differential expressions linear in the second derivatives of $g_{\mu\nu}$. The only tensors of rank two or less with this property are the contracted curvature tensor $R_{\mu\nu}$ defined by Eq. (7) and the curvature scalar $R \equiv g^{\mu\nu} R_{\mu\nu}$. Thus EINSTEIN was led (after several other attempts, including one which renounced general covariance [*32*]) to consider the equation

$$G_{\mu\nu} \equiv R_{\mu\nu} + C R g_{\mu\nu} = \varkappa T_{\mu\nu} \tag{22}$$

where C and $\varkappa$ are constants still to be determined[1].

Eqs. (22) represent ten second-order differential equations for the ten functions $g_{\mu\nu}$ and thus it appears that if $g_{\mu\nu}$ and its first derivatives are prescribed on a hypersurface $x^0 =$ constant, they are determined for all time. If this were the case, the principle of general covariance would be violated, however, since it requires that at any point the $g_{\mu\nu}$ should only be determined up to an arbitrary transformation of the four coordinates. Therefore Eqs. (22) must be such that their solutions involve four arbitrary functions, which implies that the equations must satisfy four identities [*33*]. This requirement by itself does not determine the equations. However, a particular choice of the identities allows the incorporation into Eq. (22) of a local conservation law for energy-momentum. This law must have the generally covariant form

$$T^{\mu\nu}{}_{;\nu} = 0, \tag{23}$$

which becomes a consequence of Eq. (22) provided

$$G^{\mu\nu}{}_{;\nu} = 0 \tag{24}$$

identically. This is the case only if we choose $C = -\frac{1}{2}$, or

$$G_{\mu\nu} = R_{\mu\nu} - \tfrac{1}{2} g_{\mu\nu} R; \tag{25}$$

the identities (24) satisfied by (25) are known as the contracted Bianchi identities. We thus have finally

$$R_{\mu\nu} - \tfrac{1}{2} g_{\mu\nu} R = \varkappa T_{\mu\nu}, \tag{26}$$

---

[1] For lack of time, we shall not consider here the possibility of inclusion of a "cosmological term" $\lambda g_{\mu\nu}$. There is no gain in generality in including a term proportional to $T \equiv g_{\mu\nu} T^{\mu\nu}$, since it follows by contraction of Eq. (22) with $g^{\mu\nu}$ that T is proportional to R and thus is already included, except if $C = -\frac{1}{4}$, a case which must be excluded on other grounds, as discussed later.

which can be written in the equivalent form

$$R_{\mu\nu}=\varkappa\left(g_{\mu\varrho}g_{\nu\sigma}T^{\varrho\sigma}-\tfrac{1}{2}g_{\mu\nu}g_{\varrho\sigma}T^{\varrho\sigma}\right), \tag{27}$$

where the explicit form of the right-hand side has been chosen for later convenience. By comparison with Eq. (21) in the nonrelativistic limit it is found that we must take[1]

$$\varkappa=8\pi G. \tag{28}$$

The procedure just described of incorporating the conservation law (23) into the basic equation (22) by means of identities (24) is patterned after MAXWELL's invention of the displacement current for the purpose of incorporating the law of conservation of charge into the basic equation of electrodynamics by means of an identity satisfied by these equations. The procedure is by no means necessary, however, even if a local conservation law (23) is desired. We could simply impose this law as a requirement on admissible solutions of the basic equations rather than insist on its being a consequence of the equations themselves, whose required four identities would have no connection with (23)[2].

As noted before, the space of Newtonian theory is not metric, but only affine. Thus the Γ's are not expressible in terms fo a metric tensor. They can be related to a scalar function U, however, identical with the usual gravitational potential, by

$$\Gamma^{\varrho}_{\mu\nu}=\Lambda^{\varrho}_{\mu\nu}-g_{\mu\nu}h^{\varrho\sigma}U_{;\sigma}, \tag{29}$$

where we have changed notation to denote by $\Lambda^{\varrho}_{\mu\nu}$ the affine connection used previously in Eq. (19) in the absence of a true gravitational field, i.e. the affine connection of a flat space for which

$$R^{\varkappa}{}_{\mu\lambda\nu}(\Lambda)=0. \tag{30}$$

The contracted curvature tensor formed with $\Gamma^{\varrho}_{\mu\nu}$ is given by (compare [*22, 23*])

$$R_{\mu\nu}=-g_{\mu\nu}h^{\varrho\sigma}U_{;\varrho;\sigma}, \tag{31}$$

and therefore the Poisson equation (21) can be put in the form (27) with (28), identical with EINSTEIN's equation [23]. The difference is in the requirement imposed on the solutions. In both cases the Γ's must vanish in an inertial system. But in EINSTEIN's theory we must in addition

[1] The value of $\varkappa$ depends on the choice of the components of the metric tensor $g_{\mu\nu}$ in an inertial frame. The choice (13) leads to the value (28) for $\varkappa$ which does not involve the velocity of light (compare [*23*]).

[2] Equations involving the combination $R_{\mu\nu}-\frac{1}{4}g_{\mu\nu}R$ satisfy only one identity (obtained by contraction with $g^{\mu\nu}$) and thus are not generally covariant. Such equations have been considered by EINSTEIN [*34*] and recently by PENNEY [*35*].

have Eq. (13) for the metric tensor, while in the Newtonian case we must have Eqs. (14) and (15) instead.

In the Newtonian case Eqs. (27) really contain only one independent equation and the identities (24) do not exist, since in the absence of a nonsingular metric tensor it is not possible to define $G^{\mu\nu}$ from $G_{\mu\nu}$. Therefore Eq. (23) has to be postulated separately, if a local conservation law is desired.

Our discussion has shown that EINSTEIN's postulates (C) and (E) do not determine the form of the fundamental equations uniquely. Before proceeding with an analysis of the equations chosen, we note that not only is the requirement of correspondence with the Newtonian equation (21) not sufficient, but that this requirement itself may not be necessary. Eq. (21) is only one possible form of NEWTON's law of gravitation. The more direct, original form is the one expressing the force (or equivalently the potential) explicitly as a function of the distance between the masses. This is an action-at-a-distance form, as contrasted with the (equivalent) near-action or field form (21).

Since the postulates of general relativity are local, it does indeed appear to be more natural to use a local form of Newtonian theory as a guide. However, as we shall see, general relativity can *not* dispense with global considerations, and thus EINSTEIN's postulates do not necessarily exclude the possibility that even the basic equations of general relativity might be nonlocal in character, and be expressible directly in terms of particle variables alone, in analogy to the procedure familiar from special relativity[1]. No such equations have as yet been formulated, however[2,3].

We now return to our analysis of Eqs. (27). These equations have been constructed to be generally covariant, and thus it might appear that the theory does not contain any preferred coordinate systems. It hardly needs mentioning that a class of such systems exists nevertheless in the Newtonian case, the original Newtonian inertial systems, which were only hidden by the formalism introduced before. But such systems also

[1] The simplest such formulation is to describe electromagnetic interactions in terms of Liénard-Wiechert potentials alone. A variational principle for such a theory was first given by FOKKER [*36*]; for a review of later developments see [*37*].

[2] There have been several *approximate* formulations obtained as a *consequence* of Eqs. (26) on the basis of various approximation methods. The first such formulation, applicable to slowly moving particles only, is due to J. DROSTE and H. A. LORENTZ [*38*]; a formulation applicable to fast moving particles was given by P. HAVAS and J. N. GOLDBERG [*39*].

[3] A Lorentz-invariant action-at-a-distance theory of gravitation was proposed by A. N. WHITEHEAD [*40*]; this theory and its generalization are discussed by A. SCHILD in [*9*]. The theory proposed recently by F. HOYLE and J. V. NARLIKAR [*41*], although intended as an action-at-a-distance theory, is *not* formulated in terms of particle variables alone.

exist in the case of EINSTEIN'S theory, to which we will devote ourselves exclusively from here on. First of all, the solutions of this theory have to be such that the $g_{\mu\nu}$ take on the simple values (13) in the inertial systems defined locally. Second, there is a different manner in which coordinate systems can be singled out locally, the so-called "intrinsic coordinates" [*42*]; the study of these was initiated by the need to develop a criterion to recognize which solutions of Eq. (27) are physically inequivalent, and which ones are actually equivalent, but only written in different coordinates.

While these two sets of systems are defined locally, there is a third way in which preferred reference systems arise in the theory, which is global. As we mentioned before, it had been thought by EINSTEIN that MACH'S principle is incorporated in Eqs. (27); however, it was soon realized that this was not the case[1]. To construct admissible solutions of the field equations it is always necessary (except in certain cosmological solutions which themselves embody a preferred class of reference systems) to impose boundary conditions on the metric, usually to be taken to be Minkowskian at infinity, thereby singling out a class of coordinate systems [*44*].

The three types of preferred coordinate systems just considered arise within the accepted interpretation of EINSTEIN'S theory[2]. However, there is another way in which preferred coordinate systems can be introduced, suggested independently by several authors [*46, 47*], which goes beyond this interpretation. Although this suggestion arose independently of the considerations of Newtonian theory discussed previously, it is convenient to discuss it in relation to that theory. We had found that Newtonian mechanics without gravitation can be described generally covariantly with the help of an affine connection $\Lambda^{\varrho}_{\mu\nu}$ satisfying the condition (30) for flat space, which can be used for the description of "inertial forces" due only to the use of frames of reference which are not inertial (which in our terminology implies also being Cartesian); we then introduced a second affine connection $\Gamma^{\varrho}_{\mu\nu}$ by Eq. (29) which describes both these inertial forces and "true" gravitational forces. The possibility exists of distinguishing between these two kinds of forces also in EIN-

[1] This conclusion seemed to be generally accepted until a few years ago, when the question was reopened by a different interpretation of MACH'S principle; however, examples of "anti-Mach" metrics were exhibited in [*43*]. A discussion of MACH'S principle outside the framework of EINSTEIN'S theory is given by R. H. DICKE in [*9*], p. 1.

[2] It has also frequently been argued that general relativity is not really generally covariant because of the serious restrictions imposed on the allowed coordinate transformations; this has been stressed particularly by ARZELIÈS [*27*]. For a more general discussion and critique of the concept of invariance groups with particular emphasis on general relativity see [*45*].

STEIN's theory by the use of two affine connections $\Gamma^{\varrho}_{\mu\nu}$ and $\Lambda^{\varrho}_{\mu\nu}$, the latter satisfying Eq. (30), which approach the same values far away from the sources $T_{\mu\nu}$ of the gravitational field. This in itself is sufficient to tie EINSTEIN's theory to a flat space, and whether we then wish to interpret it as a flat space or a curved space theory becomes a question of semantics; exactly as in the case of the Newtonian theory, it depends only on which affinity we want to consider as the affinity of the "true" space.

In such a theory of two affine connections the connection $\Gamma^{\varrho}_{\mu\nu}$ is as usual the Christoffel symbol (10) formed from the metric tensor $g_{\mu\nu}$. But in addition a second nonsingular tensor $\lambda_{\mu\nu}$ can be introduced such that the Christoffel symbol formed with $\lambda_{\mu\nu}$ gives the connection $\Lambda^{\varrho}_{\mu\nu}$. This tensor furnishes a second fundamental quadratic form $d\sigma^2$ in addition to ds^2 given by Eq. (9):

$$d\sigma^2 = \lambda_{\mu\nu} dx^{\mu} dx^{\nu}. \tag{32}$$

The introduction of these quantities can be looked at simply as a formal device which leads to a formulation of the equations of EINSTEIN's theory more convenient for certain purposes than the customary one [*46*]. But beyond this possibility of a formal interpretation of EINSTEIN's theory as a flat-space theory, another question was raised by ROSEN [*46*]. As an alternative to the purely formal introduction of the tensor $\lambda_{\mu\nu}$ he suggested the possibility of considering this tensor rather than $g_{\mu\nu}$ as the actual metric of the physical space-time, and to identify it (in a suitable class of coordinate systems) with the Minkowski metric (13). This amounts to a different assumption of the behavior of clocks and measuring rods than that adopted in EINSTEIN's theory. The problems connected with this interpretation are discussed in detail by ROSEN.

In the formulation of Newtonian theory discussed earlier the connection $\Lambda^{\varrho}_{\mu\nu}$ is indispensable, but there is no need to relate it to a tensor $\lambda_{\mu\nu}$. However, we are free to do so if we desire to carry the analogy with ROSEN's formalism as far as possible.

As far as the structure of both theories is concerned, $\lambda_{\mu\nu}$ is completely arbitrary except for the requirement of flatness (30). We therefore might be tempted to try to imitate ROSEN's suggestion and to use $\lambda_{\mu\nu}$ to introduce a Minkowski metric into Newtonian theory. Formally this is indeed possible; however, any attempt to interpret it as the actual metric of space-time must fail, since the physical conditions imposed on signals by the Minkowski metric and by the Newtonian space-time concepts are incompatible.

As we discussed before, it was basic for the interpretation of the theory of general relativity that the equation of motion of a particle (without internal structure) should be a geodesic in the absence of nongravitational fields. Nevertheless, from our arguments it appeared that the geodesic law (19) had to be postulated independently of the field equations, and this was indeed the procedure adopted by Einstein originally. However, it was soon realized by various authors that the equations of motion are not independent of the field equations. Numerous derivations of the geodesic law were given (some as early as 1921), and it was further realized that the geodesic law was a consequence of the covariant conservation law (23) for the energy-momentum tensor *alone* [*48*]. Moreover, *any* generally covariant theory with a law (23) contains the geodesic law (19), even if the space-time of such a theory is only affine, but not metric; in particular, this is the case for the Newtonian theory considered before [*49*]. Thus both EINSTEIN'S and NEWTON'S theory contain the law of motion for a particle as a consequence of the conservation law (23). This conservation law in turn is a consequence of the field equations in EINSTEIN'S theory.

Although it might appear that it is a major advantage of EINSTEIN'S theory that it automatically entails the conservation of energy-momentum and of the equations of motion, it is precisely these features which are coupled with two major difficulties of the theory. The common origin of both the advantages and the difficulties is the fact that in Eq. (27) the energy-momentum tensor must include *all* nongravitational fields as well as *all* matter[1]. Therefore the theory cannot deal with open systems, but only with closed ones, strictly speaking only with the universe as a whole [*50*]. This is in contrast not only to Newtonian theory, but also to MAXWELL'S equations, in spite of the fact that the latter contain an identity just as Eq. (27) does. This feature leads to serious difficulties in the treatment of the problem of transmission of energy and information, and thus of gravitational radiation, as the usual methods of studying the effects on a field of an arbitrary motion of the sources no longer apply; it has therefore been suggested [*51*] that it will be essential to study nonanalytic solutions of Eqs. (27), but the problems of energy and radiation are far from settled [*52*].

The question of the energy of a gravitational field has plagued the general theory of relativity from its inception. We noted earlier that seemingly Eq. (27) contained all the energy sources *except* the gravitational field. However, the covariant conservation law (23) is not of the form of an ordinary divergence (as customary for such a law formulated in a Cartesian inertial frame of reference and as needed to permit the formulation of a global conservation law by integration) except in an inertial system. As the gravitational field is described by the affine connections,

[1] In Newtonian theory there is no contribution by nongravitational fields from $T^{\varrho\sigma}$ in Eq. (27), whether or not they are formally included [*23*].

one would expect from analogy with special relativistic field energies that the expression for the energy-momentum of the gravitational field should be quadratic in these quantities; but then it vanishes in any local inertial system and thus can not be a tensor. Such an expression (a "pseudotensor") was suggested by EINSTEIN, who showed that it allowed the transformation of Eq. (23) into an ordinary divergence and the interpretation of Eq. (26) as containing also gravitational energy as a source of gravitation. Nevertheless, there are a number of difficulties associated with EINSTEIN's definition as well as with the alternative definitions suggested by other authors [*52*, *53*].

The very serious difficulties still connected with the concept of gravitational energy in EINSTEIN's theory after half a century of exploration make it appear rather doubtful that there is any advantage to *basing* this theory on this very concept. Such a procedure has been suggested repeatedly, taking as a starting point a Lorentz-invariant theory in which all forms of field energy including that of gravitation are taken to be sources of gravitation [*54*]. This is all the more questionable as the concept of localized field energy is not free of difficulties even in special relativistic theories. Moreover, the form of the energy-momentum tensors is not fully determined within the framework of such theories, but appeal has been made to the *general* theory of relativity to find the proper form of the special relativistic tensors [*55*].

One of the major problems we encountered in our discussion of energy and its transmission was that of the necessity of treating EINSTEIN's theory as based on Eq. (26) as describing a closed system. This has to be contrasted with our general discussion of properties of space-time based on EINSTEIN's two postulates only. The question of open or closed systems did not arise there, since no connection between the properties of space-time and the sources responsible for them had yet been introduced. These properties were considered to be explorable by means of macroscopic physical instruments. It has frequently been asserted that EINSTEIN's theory itself only applies to macroscopic phenomena and should not be applied on the microscopic scale. I shall not discuss here the arguments for and against this point of view; the final decision between them will have to be by experiment. I shall only touch on a few consequences of one or the other attitude.

If the theory is taken to be applicable to arbitrarily small scale phenomena, first of all there is physical meaning to be attributed to treating space-time as a differentiable manifold [*56*] and to solutions of Eqs. (26) over the full range of the coordinates. On the one hand this forces us to accept any possible difficulties (such as singularities) inherent in such solutions as physically significant; on the other, it allows us to go beyond what would be considered acceptable macroscopic solutions by considering e.g. non-Euclidean topologies of space-time in the small [*30, 31*], or non-Minkowskian signatures of the metric in the small [*57*]. Second, the microscopic description of a closed system has to be related to the macroscopic measurements of everyday physics, a problem not satisfactorily solved even in Newtonian physics or in special relativity, where geometric quantities are not dynamical variables. Third, since it is clear that classical physics is not sufficient for the description of atomic or subatomic phenomena, it becomes imperative to reconcile general relativity with quantum theory, a task which would be enormously complicated if geometric quantities would have to be quantized [*58*].

If the theory is taken to be applicable on the macroscopic scale only, the difficulties just discussed would be considered as pseudoproblems. On the other hand, one would renounce the possibility of making use of the new mathematical possibilities offered by EINSTEIN'S equations for the understanding of elementary structures.

We have discussed a number of problems basic to the formulation of EINSTEIN'S general theory of relativity. Obviously it is impossible to cover all the questions raised by the innumerable friendly or hostile students of this theory, and any selection must be highly subjective. In the selection offered here a number of questions had to be left unanswered. These have also been left open by all of the theories offered until now which are extensions or generalizations of EINSTEIN'S theory [*59*]; all of these theories share the mathematical and epistemological features of EINSTEIN'S theory which we discussed and the new physical ideas incorporated into some of them do not remove any of the particular difficulties mentioned, while some add a few of their own.

The view once widely held was that "the great triumph of the theory of relativity lies in its absorbing the universal force of gravitation into the geometrical structure; its success in accounting for minute discrepancies in the Newtonian description of the motions of test-bodies in the solar field, although gratifying, is nevertheless of far less moment to the philosophy of physical science. EINSTEIN'S achievements would be substantially as great even though it were not for these observational tests" [*60*]. Half a century of study of the theory by two generations of physicists and mathematicians has indeed left us with the same admiration of EINSTEIN'S achievements. Nevertheless, we can no longer fully agree with this view. On the one hand, the possibility of unifying gravitation and geometry, as we have seen, exists even in Newtonian theory. On the other, the possibility exists to force Lorentz invariant theories of gravitation to yield the same observational effects as EINSTEIN'S theory, without directly linking gravitation and geometry. The overpowering attraction of EINSTEIN'S theory lies in its ability to provide this link as well as the observational tests in a natural manner;

but like any other physical theory it must stand or fall on its agreement with experiment. If future tests should show a disagreement (and we can be certain from the history of science that some day a discrepancy will be found, or the inadequacy of the theory for some phenomena will be established), we might have to abandon EINSTEIN's field equations. However, from our discussion it is clear that there are other theories which fit into the framework of EINSTEIN's two postulates, and thus such a disagreement would not necessarily be fatal to general relativity. The study of such alternate theories is interesting for its own sake, but it will not be their philosophical impact which will determine whether one of them, or indeed a theory which is not based on EINSTEIN's postulates, will supersede EINSTEIN's theory, but only their ability to provide a better description of nature than is provided by the present theory.

REFERENCES

[*1*] For a review of the early attempts see M. ABRAHAM, Jahrb. Radioakt. u. Elektronik **11**, 470 (1914).

[*2*] See e.g. C. MØLLER, The theory of relativity; Oxford: Oxford University Press 1952, or D. BOHM, The special theory of relativity; New York: W. A. Benjamin, Inc. 1965.

[*3*] KLEIN, F.: Vorlesungen über die Entwicklung der Mathematik im 19. Jahrhundert, Bd. 2, Kap. 2. Berlin: Springer 1927.

[*4*] The space-time structure of the special theory of relativity is discussed in all textbooks of this theory; a particularly detailed account is given in H. ARZELIÈS, La cinématique relativiste. Paris: Gauthier-Villars 1955.

[*5*] For a recent review of the special theory see the article on special relativity by P. G. BERGMANN in Encyclopedia of physics, vol. IV, p. 109. Berlin-Göttingen-Heidelberg: Springer 1962.

[*6*] For further discussion of this point see [*5*] and P. HAVAS, Relativity and causality. In: Proceedings of the 1964 Internat. Congr. for Logic, Methodology and Philosophy of Science. p. 341. Amsterdam: North-Holland Publ. Co. 1965.

[*7*] A survey of recent such theories is given by G. J. WHITROW and G. E. MORDUCH, Nature **188**, 790 (1960) and in: Vistas in Astronomy ed. by A. BEER, vol. 6, p. 1. New York: Pergamon Press 1965); also P. KUSTAANHEIMO and R. LEHTI, Soc. Sci. Fennica Commentationes Phys.-Math. **28**, No. 2 (1963).

[*8*] The old attempts of explanation are discussed in J. CHAZY, La théorie de la relativité et la mécanique céleste, vol. I, Chap. V. Paris: Gauthier-Villars 1928.

[*9*] Evidence for gravitational theories (Proceedings of the Internat. School of Physics "Enrico Fermi", Course XX, ed. by C. MØLLER). New York and London: Academic Press 1962.

[*10*] The development of EINSTEIN's ideas has been described by him in his "Autobiographical notes" in ALBERT EINSTEIN: Philosopher-scientist, ed. by P. A. SCHILPP, p. 1. New York: Tudor Publ. Co. 1949.

[11] MACH, E.: Die Mechanik in ihrer Entwicklung. Leipzig: F. A. Brockhaus 1885.

[12] JAMMER, M.: Concepts of space. Cambridge: Harvard University Press 1954.

[13] ZANSTRA, H.: Phys. Rev. **23**, 528 (1924).

[14] EINSTEIN, A.: Ann. Physik **49**, 769 (1916).

[15] EÖTVÖS, R.: Gesammelte Arbeiten. Budapest: Akadémiai Kiadó 1953.

[16] EINSTEIN, A.: Ann. Physik **35**, 898 (1911).

[17] KRETSCHMANN, E.: Ann. Physik **53**, 575 (1917).

[18] EINSTEIN, A.: Ann. Physik **55**, 241 (1918).

[19] Nevertheless, most expositions of general relativity fail to point out the lack of physical content of the principle of general covariance. Notable exceptions are L. SILBERSTEIN, The theory of general relativity and gravitation (New York: Van Nostrand Co. 1922), R. C. TOLMAN, Relativity, thermodynamics, and cosmology (Oxford: Oxford University Press 1934), and especially V. FOCK, The theory of space time and gravitation (English translation: New York: Pergamon Press 1959).

[20] Encyclopedia of physics, vol. III/1. Berlin-Göttingen-Heidelberg: Springer 1960.

[21] CARTAN, E.: Ann. Ec. Norm. **40**, 325 (1923); **41**, 1 (1924), reprinted in Oeuvres complètes, vol. III/1, pp. 659 and 789. Paris: Gauthier-Villars 1955.

[22] The case of gravitational fields without sources was treated by E. CARTAN [21] in a formulation somewhat different from that of EINSTEIN's theory, and by K. FRIEDRICHS, Math. Ann. **98**, 566 (1927) in a formulation very close to EINSTEIN'S.

[23] A detailed exposition of this problem is given in P. HAVAS, Revs Mod. Phys. **36**, 938 (1964), which also contains a treatment of the gravitational field with sources generalizing FRIEDRICH's formulation.

[24] Such a redefinition of inertial systems was suggested first by O. HECKMANN and E. SCHÜCKING, Z. Astrophys. **38**, 95 (1955), in connection with considerations on Newtonian cosmology.

[25] For a brief introduction well suited for our purposes see E. SCHRÖDINGER, Space-time structure; Cambridge: Cambridge University Press 1950. We shall only use the traditional mathematical apparatus of general relativity here. For a formulation of the theory by means of a "direct analysis" rather than explicit use of coordinates see J. A. SCHOUTEN, Verhandel. Koninkl. Akad. Wetenschap. Amsterdam **12**, No. 6 (1918). A treatment using modern mathematical methods is given by J. M. SOURIAU, Géometrie et relativité. Paris: Hermann 1964.

[26] MINKOWSKI, H.: Nachr. Akad. Wiss. Göttingen, Math.-physik. Kl. 53 (1908); Math. Ann. **68**, 472 (1910).

[27] For a particularly detailed discussion see C. MØLLER [2] and H. ARZELIÈS, Relativité généralisée. Gravitation, fasc. 1. Paris: Gauthier-Villars 1961.

[28] TONNELAT, M.-A.: La théorie du champ unifié d'EINSTEIN. Paris: Gauthier-Villars 1955.

[29] HLAVATÝ, V.: Geometry of EINSTEIN's unified field theory. Groningen: P. Noordhoff Ltd. 1957.

[30] EINSTEIN, A., and N. ROSEN: Phys. Rev. **48**, 73 (1935).

[31] WHEELER, J. A.: Geometrodynamics. New York and London: Academic Press 1962.

[32] EINSTEIN, A., and M. GRASSMANN: Z. Math. u. Phys. **62**, 225 (1913).

[33] HILBERT, D.: Nachr. Akad. Wiss. Göttingen, Math.-physik. Kl. 395 (1915).

[34] EINSTEIN, A.: Sitzber. preuss. Akad. Wiss., Physik.-math. Kl. 349 (1919).

[35] PENNEY, R.: Phys. Rev. **137**, B 1385 (1965).

[36] FOKKER, A. D.: Z. Physik **58**, 385 (1929).

[37] HAVAS, P.: Classical relativistic theory of elementary particles. In: Argonne national laboratory summer lectures on theoretical physics, 1958 (ANL-5982, 1959), p. 124.

[38] DROSTE, J., and H. A. LORENTZ: Versl. Kon. Akad. Wet. Amsterdam **26**, 392 (1917), reprinted in H. A. LORENTZ, Collected papers, vol. V, p. 330. The Hague: M. Nijhoff 1937.

[39] HAVAS, P., and J. N. GOLDBERG: Phys. Rev. **128**, 398 (1962).

[40] WHITEHEAD, A. N.: The principle of relativity. Cambridge: Cambridge University Press 1922.

[41] HOYLE, F., and J. V. NARLIKAR: Proc. Roy. Soc. (London) A **282**, 184, 191 (1964) and A **294**, 138 (1966). For a critique see S. DESER and F. A. E. PIRANI: Proc. Roy. Soc. (London) A **288**, 133 (1965).

[42] For a review of this problem see the article on general relativity by P. G. BERGMANN in [5], p. 203.

[43] OZSVÁTH, I., and E. SCHÜCKING: In: Recent developments in general relativity, p. 339. New York and Warszawa: Pergamon Press-PWN 1962, and Nature **196**, 363 (1962); see also H. HÖNL and H. DEHNEN: Z. Physik **191**, 313 (1966) for a different view, and references given in these papers.

[44] For a recent brief discussion and references to the literature see A. GRÜNBAUM, Philosophical problems of space and time, Ch. 14. New York: Alfred A. Knopf 1963. Compare also V. FOCK, [19] and Revs. Mod. Phys. **29**, 325 (1957).

[45] BERGMANN, P. G.: In: Fundamental topics in relativistic fluid mechanics and magnetohydrodynamics, ed. by R. WASSERMANN and C. P. WELLS, p. 29. New York and London: Academic Press 1963.

[46] ROSEN, N.: Phys. Rev. **57**, 147, 150 (1940) and Ann. Phys. (New York) **22**, 1 (1963). For some critical comments see W. BAND, Phys. Rev. **61**, 698 (1942).

[47] PAPAPETROU, A.: Proc. Irish Acad. **52**, 11 (1948); KOHLER, M.: Z. Physik **131**, 571 (1952); **134**, 286, 306 (1953); KRAUS, K.: Z. Physik **168**, 61 (1962).

[48] A simple derivation and references to the literature are given by HAVAS and GOLDBERG [39]. The early history of the problem is discussed by P. HAVAS in a paper to be published shortly.

[49] HAVAS, P.: J. Math. Phys. **5**, 373 (1964). The generalization to the case of the presence of nongravitational fields is given in [23].

[50] Some of the physical differences in the treatment of open and closed systems in the theory of relativity (mainly in the special one) are discussed in [6].

[51] BONDI, H., M. G. J. VAN DER BURG, and A. W. K. METZNER: Proc. Roy. Soc. (London) A **269**, 21 (1962).

[52] See the reports and discussion at the Internat. Meeting in Florence 1964 on "Problems of energy and gravitational waves" in: Pubblicazioni del comitato nazionale per le manifestazione celebrative del IV centenario della nascita de Galileo Galilei, Vol. II, tomo 1. Firenze: G. Barbèra 1965.

[53] For a review see A. TRAUTMAN in Gravitation, ed. by L. WITTEN, ch. 5. New York: John Wiley & Sons, Inc. 1962.

[*54*] GUPTA, S.: Revs. Mod. Phys. **29**, 334 (1957); THIRRING, W.: Fortschr. Physik **7**, 79 (1959); Ann. Phys. (New York) **16**, 96 (1961). For further references and a critique of this approach see L. HALPERN, Ann. Phys. (New York) **25**, 387 (1963).

[*55*] ROSENFELD, L.: Mém. Acad. roy. belg. **18**, No. 6 (1940); BELINFANTE, F. J.: Physica **7**, 449 (1940).

[*56*] The possible need for abandoning this assumption is discussed in the paper on "Physics and geometry" by P. G. BERGMANN in [*6*].

[*57*] TREDER, H.: Ann. Physik **9**, 283 (1962); LANCZOS, C.: J. Math. Phys. **4**, 951 (1963).

[*58*] This is by no means necessary, however, although almost universally implied. The contrary view is concisely argued by L. ROSENFELD, Nuclear Phys. **40**, 353 (1963). For a survey of some recent work on quantization and the problems involved see [*42*] and B. DE WITT in [*53*], ch. 8.

[*59*] A general "theory of field theories" has been developed by D. G. B. EDELEN, and summarized in The structure of field space (Berkeley and Los Angeles: University of California Press 1962), which also contains many references to unified field theories. Compare also [*28*], [*29*] and D. R. BRILL in [*9*], p. 51.

[*60*] ROBERTSON, H. P.: In: [*10*], p. 329.

Chapter 9

Relations of Quantum to Classical Physics[1]

Ralph Schiller

Department of Physics, Stevens Institute of Technology
Hoboken, New Jersey

I

If, as Einstein believed, nature were not possessed of malice, then surely she would have created matter otherwise. For it is either malice or capricious ambivalence on her part which has brought us to our present state where we can say with great authority that matter is not this or that, and yet not utter a single unqualified phrase as to what it might be.

I seem to have outrun my tale and I must really commence at some point closer to its beginning. In the early part of this century there was a prevailing view as to matter's nature, and this picture we call classical. As we know this view held that matter consisted of two related, but nevertheless distinct elements, idealized particles and fields. The idealized particle was conceived to be a point-like structure which could move through space, and in particular could be accelerated if acted on by an external field. The particle's reaction to the field was governed by Newton's laws of motion. A given particle's behaviour depended on the external field, and also on two parameters, its mass and charge. These parameters identified the particle and their values were presumed independent of the external fields.

Macroscopic matter consisted of many such idealized particles bound together by mutual interaction. The precise behaviour of a given macroscopic body in arbitrary external fields could be quite complex because its motion was determined by the underlying motions of all the elementary particles in the macroscopic object. Only bodies with extremely simple properties, e.g., rigidity, symmetry, etc., lent themselves to immediate analysis, and even these simplifications proved inadequate for exact analysis when the external fields became intricate. For simple fields however, the laws of motion of these idealized macroscopic bodies

[1] This work was supported in part by the Office of Aerospace Research, USAF.

proved to be quite similar to the laws of motion originally postulated for point particles. The motion again depended on the total mass and charge of the massive object, although the description of the additional degrees of freedom associated with rotation (point particles were assumed to be true, non-oriented points and lacked these degrees of freedom) depended on knowledge of additional parameters such as moments of inertia, dipole moments, etc. As the complexity of the external fields increased more parameters describing the body were required. The description of the motion became increasingly difficult, especially since these parameters were in general complicated functions of the time.

And yet despite these formal difficulties the set of ideas was complete; for with knowledge of the forces (fields) acting between idealized particles one could calculate, at least in principle, the motion on any macroscopic object.

Such was the classical view of the particle at the turn of the century. However it was not an immutable set of ideas. For example, since the point particles were idealized conceptual elements, there was no reason why a direction (orientation, or in modern parlance, polarization) could not be associated with each point. In fact the particle could be endowed with an unlimited number of internal characteristics, and the laws of motion describing the time variation of these internal parameters could also be postulated. The particles could then be placed in various fields, or permitted to interact with one another, to uncover whether these novel qualities did exist in fact as well as in fancy. Further, macroscopic matter consisting of such elements would not have the same properties as matter constructed from our original point particles. Analysis could determine these properties and comparison made with experimental observations on real objects.

Not only was such a program feasible in those days, but it was also realized by the early 1900's. And if one supposes that these investigations lay outside the main thrust of the physics of that period, one could not be farther from the truth. On the contrary, they were part and parcel of the great attempt (that it ended in failure is irrelevant) to comprehend the nature of electromagnetic processes[1].

I have detailed this arrested line of development for classical physics, certain that it is only one of many, in order to show that the present prevailing view of classical physics as a closed subject is unwarranted. My purpose of course is not to defend the honor of classical physics, but rather to ask whether these and other conceptual possibilities have relevance for modern quantum theory. My thesis is that they do, although I am not of the opinion that quantum theory can be reduced to

[1] Many of the so-called aether theories were particle theories of this type.

traditional classical physics. The relevance of which I speak has to do with the visualization and comprehension of phenomena; both denied us in many ways and often without cause in the quantum theory. Again I seem to have outrun my tale, so let us return.

I have spoken in some detail about particles, neglecting to a major extent the other half of the classical coin, the fields. These were presumed continuous everywhere in space except at the position of particles where the field intensity became infinite. The fields obeyed specific laws and the best known were LAPLACE's equation and MAXWELL's equations for the gravitational field and electromagnetic field, respectively. As in the case of idealized particles the scheme was not immutable and richer field structures could be postulated, and indeed were. The classic example was EINSTEIN's gravitational field which replaced the field of LAPLACE.

The introduction of EINSTEIN's gravitational field, a far more significant change than his earlier special theory of relativity, profoundly altered our view of matter as well as of space-time. In the theory of gravitation the bond between the particle and the field was drawn closer, but the particle still evidenced itself in a region where the field was undefined. Thus in the general sense classical matter remained field and particle — related but distinct.

There were attempts over the years to achieve a unified view of matter by reducing the field to the particle concept, or the particle to the field, but none seemed to meet with any real success. Significant efforts which come to mind are the electrodynamic theories of MIE, and of BORN and INFELD, which sought to reduce the particle to a form of concentrated, but not infinite, field energy. In the other direction we had the electromagnetic theory of WHEELER and FEYNMAN which dispensed entirely with the field.

The apparent failure of these efforts did not resolve the issue. With reluctance, physicists again considered particle and field as related but nevertheless distinct entities. The hope was that eventually a unified view would be achieved; a hope as yet unfulfilled.

II

We all know that the greatest failure of classical theory lay in its complete inability to cope with atomic structure, while SCHRÖDINGER's wave theory of matter had spectacular success in explaining this structure. SCHRÖDINGER's original view was that his equation described a classical matter field, so that the true motion of particles was similar to the behaviour of the waves of classical optics. If this view had held classical concepts could have been retained with the classical particle replaced by the classical field. However the interpretation failed, and with its failure came the general breakdown of the classical view of matter. The reasons for this failure form a part of our tale.

All classical fields that change in time exhibit a periodic behaviour which we describe by the general term, "wave." It is also a fact that all known wave-like phenomena possess a particle-like (ray) limit in much the same way that geometrical (ray) optics is the limit of classical wave optics. If one considers high frequency waves and avoids spatial regions near sharp boundaries, then the mathematical and pictorial representation of advancing wave fronts may be transformed into an analysis of the motion of rays perpendicular to these wave fronts. These rays have a particle-like aspect although the approximate, sometimes called asymptotic, wave theory refers only to an ensemble of rays, i.e., to a continuous distribution of rays (a wavefront) and not to individual rays. On the other hand the general formalism of the ray theory always permits one to choose one of the rays from the ensemble and observe its motion to the exclusion of all the others. The apparent ability to trace single rays in the theory is mirrored in our capacity to focus very narrow beams of light or sound of high frequency.

We thus have the impression that we are maneuvering the single rays of our theory. However individual rays are a fiction, for practice tells us that we have misused our ray theory. The continuity of both the wave front and the distribution of rays is real, since the more we narrow our beam the more it will spread (diffract). Thus we can never evidence the individual ray. In reality, only an ensemble of rays exists, and in the above example the ensemble will eventually manifest itself in the spreading of the beam.

These properties of an ensemble of rays can be characterized concisely by means of an uncertainty principle.

$$\Delta x \, \Delta p_x \simeq p_x / k \tag{1}$$

where Δx is the uncertainty of the beam width, $\Delta p_x / p_x$ the uncertainty in the beam (ray) direction orthogonal to the beam itself, and k is the wave number $2\pi/\lambda$. We emphasize that this uncertainty principle is characteristic of all known classical wave phenomena. The uncertainty principle valid for classical matter waves is a consequence of the general law (*1*) and the restriction of the momentum values to $p_x = \hbar k$. It takes the form

$$\Delta x \, \Delta p_x \simeq \hbar , \tag{2}$$

where $\hbar$ is PLANCK'S constant.

The fundamental description of any wave phenomenon is contained in the exact solutions of its wave equation. However under a variety of circumstances a great deal of information regarding the structure of the wave may be gleaned from an analysis of individual ray motions. From these ray trajectories we are led to ensembles of rays, and thus to approximate wave fronts in regions where diffractive effects are unimportant. If we require that our wave fronts be continuous, bounded, and single-valued functions of position, we can construct in several cases exact solutions, and in general very reasonable approximate solutions, to the original wave equations. The process of constructing the wave solutions from the ray motions is an outgrowth of a general procedure called the JWKB approximation, after JEFFREYS, WENTZEL, KRAMERS, and BRILLOUIN.

However one should not be misguided by the success of this method into presuming that we have replaced the wave by its ray aspect, or in a more specific context reduced quantum mechanics to classical dynamics. On the contrary, we really do not have a wave until we require that the wave fronts constructed from the rays behave as waves, i.e., satisfy conditions of contuinity, boundedness, single-valuedness, etc. It is just these requirements which distinguish ensembles of rays from true waves.

How wide is the range of validity of the JWKB approximation? We know for example that classical optics is an approximation to the full set of MAXWELL'S equations, and that the Schrödinger wave equation is only an approximation to the PAULI wave equation[1]. In these cases, and in others as well, the differences between equations, e.g., between the scalar wave equation and MAXWELL'S equations, is mirrored in the character of the rays associated with the asymptotic wave fronts. For the scalar wave of classical optics we have rays without internal degrees of freedom, while there is an intrinsic orientation associated with the rays of MAXWELL'S equations. In fact in the ray limit of MAXWELL'S theory we find two sets of equations describing the ray [*1*]. One of them yields its trajectory in space; the other describes the continuous change in the internal state of polarization as we move along the ray path. These properties of rays with internal degrees of freedom represent a general feature of any wave phenomenon where the wave fields is geometrically more complex than a scalar function. And again as in the earlier cited cases of scalar wave fields, the utilization of the solutions of all the ray equations, including the polarization equation, provides us with a versatile method for finding approximate, and sometimes exact, solutions of the original wave equation.

In view of this very general state of affairs in wave theories, it may prove a distinct surprise to learn that many physicists hold a contrary

---

[1] We are neglecting relativistic wave theories of massive particles, although statements similar to those which follow way be made in regard to those theories.

opinion, at least for the quantum theory, and subscribe to the following thesis: "In contrast to the charge and mass of the electron, the spin cannot be said to belong to the classically defined properties of the atomic model." [2] The above denial of classical (ray) significance to the intrinsic polarization of the electron is due to BOHR and is based on the following principle: There is no way of verifying the existence of spin by means of a Stern-Gerlach experiment. BOHR'S comment is absolutely true, and yet it fails to contradict our general thesis which holds that in those regions of space where the ray picture is applicable, spin is a classically describable concept. Where do we differ? A Stern-Gerlach experiment maximizes the effects of diffraction, and it is precisely under such conditions that one would not expect the spin to be a valid classical (ray) concept. On the other hand there are many situations where diffraction effects are minimized, and in these circumstances we expect that, to a high degree of accuracy, the spin will obey classical equations. The truth of this matter is assured by the numerous experiments in recent years with beams of charged, polarized particles which are adequately described by the ray equations for trajectory and spin [3]. In ruling out spin as a classical entity BOHR made an unfortunate choice of experimental arrangement; for on similar grounds we could equally well rule out the useful concept of momentum as a vector by requiring that the momentum of a particle always be determined by means of experiments with beams passing through very narrow slits. Under those conditions momentum as a directed quantity also loses its significance.

We have been discussing certain general characteristics of wave phenomena and now we must examine in greater detail the matter waves of SCHRÖDINGER. In particular we ask whether the particles that we observe in nature are the rays associated with SCHRÖDINGER'S matter waves. The answer is no! For while mass particles, e.g., electrons, generally follow along ray paths, they are totally unlike rays in that under diffraction they do not divide into smaller parts. Thus, if we treat the SCHRÖDINGER equation as a classical wave equation, we cannot predict the existence of individual particles, and there is no simple way of justifying the oft made claim that in some limit the quantum theory goes over to the classical physics of a point particle. And yet the Schrödinger equation treated as a second quantized field in some sense describes both a wave and a particle, so that it is not simply an equation for matter waves. What then is the Schrödinger equation, and what does it describe?

This question touched off the most significant debate of this century concerning the nature of matter; a debate that apparently has not yet come to a close. One group of physicists, the Copenhagen School, answered the question as follows: The existence of objects which simultaneously possess particle-like and wave-like properties shows that we must renounce forever the possibility of forming any pictorial image of microscopic events. Such events are closed off from view since they take place in a fog enshrouded world whose boundaries form the only identifiable objects. This interpretation of atomic events holds that unlike other physical processes studied in the past, microscopic phenomena cannot be defined independently of the macroscopic measuring instruments used in a given experiment. These instruments obey classical laws, laws necessarily independent of the laws of the quantum theory, and they alone provide us with the sense perceptions that are meaningful in creating mental images of quantum phenomena. These images are the idealized point particles and the fields of classical physics, and they form the boundaries (or extremes) of our mental picture. To my knowledge no other visual element has ever entered into this interpretation of the quantum theory.

This revolutionary view of matter has proved a consistent attitude over the years, and in this sense has been adequate for our comprehension of atomic phenomena. It is a view of nature subscribed to by most physicists; hence our opening comment.

However this attitude has been sharply attacked by many prominent scientists who have held that the Copenhagen interpretation was an inadequate base on which to rest a general view of natural phenomena. They suggested that one should not forego the possibility of forming some definite hypothesis concerning the nature of microscopic processes, no matter how formidable such a task might prove. Many attempts were made over the years to create a viable alternative to the Copenhagen interpretation, yet most of these efforts failed to gain widespread acceptance among physicists. I shall not provide an assessment of these efforts since that would take us too far afield. Instead I should like to pursue a line of thought initiated by SCHRÖDINGER [*4*]; a view far different from his original interpretation of the quantum theory as a theory of matter waves.

III

The paradox of the quantum theory lies in the simultaneous existence of particle and wave in a single phenomenon. Experiments have shown that individual particles obey the diffraction rules for waves; for they appear only in those regions of space where the wave amplitude does not vanish. The fundamental issue in resolving this paradox is whether or not "it is obvious that a thing cannot be a form of wave motion and composed of particles at the same time — the two concepts are too different" [*5*]. I believe that such an amalgam is not beyond the realm of comprehension. In fact classical entities may be formed with just this property, viz., simultaneously exhibiting wave and particle properties. Furthermore they are not esoteric constructs unrelated to the phenomena observed. To the contrary, some of these classical quantities are exact solutions of the equations of the second quantized field, the quantum theory which unites the wave and particle. At worst, these mathematical entities are meaningful approximate solutions in the sense of the JWBK approximation to these same equations. I shall try to describe these entities through an illustrative example.

The ray ensemble limit of the Schrödinger equation is the classical theory embodied in the Hamilton-Jacobi equation

$$\frac{\partial S}{\partial t} + H(\nabla S, x, t) = 0. \tag{3}$$

The function S defines a surface which "propagates" in time. The same S is the phase of the matter wave in regions where diffraction effects are not important. The rays are orthogonal to the surface S = constant. And since we expect that rays in the ensemble are neither created nor destroyed, it is not surprising that S defines another quantity D which satisfies an equation of continuity

$$\frac{\partial D}{\partial t} + \nabla \cdot (D\,\partial H/\partial \nabla S) = 0. \tag{4}$$

D represents the density of ensemble particles, and Eq. (4) is a conservation law for the rays of the ensemble. Both Eqs. (3) and (4), are derivable from a variational principle where the action integral is [6]

$$\Sigma = \int d^4x \{D[\partial S/\partial t + H(\nabla S, x, t)]\} \tag{5}$$

and the independent variables in Σ which are to be varied are the functions S and D[1]. If we now make the transition to the Hamilton theory of (5), we find that the momentum density canonically conjugate to the variable S is $-D$,

$$\pi = \delta\Sigma/\delta\dot{S} = -D. \tag{6}$$

π and S satisfy the usual Poisson bracket relation

$$-[\int \pi(x')\, d^3x', S(x)] = 1. \tag{7}$$

The Hamiltonian of our classical theory is

$$\overline{H} = -\int d^3x [\pi H(\nabla S, x, t)]. \tag{8}$$

[1] A more appropriate "canonical" variable would be e^{iS} in place of S. Its introduction would not affect the argument in this paper.

The transition from this classical Hamiltonian to a quantum field theory is effected in the usual way,

$$\bar{H}_{\mathrm{op}} \Psi = i\hbar\, \partial\Psi/\partial t, \tag{9}$$

where $\bar{H}_{\mathrm{op}}$ is the classical $\bar{H}$ of (8) with π replaced in some symmetric fashion by the operator $-i\hbar\,\delta/\delta S$. Ψ is some functional of S, x and t.

Equation (9) admits the following constant of the motion, $N_{\mathrm{op}} = -\int \pi\, d^3 x$,

$$(N_{\mathrm{op}} \bar{H}_{\mathrm{op}} - \bar{H}_{\mathrm{op}} N_{\mathrm{op}}) \Psi = 0. \tag{10}$$

N_{op} is the particle number operator which corresponds to the total number of rays in the classical theory. Equation (10) shows that Ψ can be a simultaneous eigenstate of N_{op} and $\bar{H}_{\mathrm{op}}$,

$$N_{\mathrm{op}} \Psi = N \Psi. \tag{11}$$

N is a number and its possible values depend on additional requirements on Ψ. The simplest demands on the "wave functional" Ψ lead to integral values of N, $N = 1, 2, 3, \ldots \infty$, although other requirements are possible and these lead to different values of N. If $N = 1$, we say that one "particle" is present. However this is no classical point particle we have found, although at times it might appear so.

The classical theory associated with the "wave equation" (9) is now more general than our original classical Hamilton-Jacobi equation (3) and the derivative equation (4). The new classical theory is now an ensemble theory of "rays", or more accurately, an ensemble theory of ensembles. The new classical theory is the Hamilton-Jacobi theory implicit in (5) and (8),

$$\partial\Sigma/\partial t + \bar{H}([\delta\Sigma/\delta S], [S]) = 0. \tag{12}$$

The claim now is that Ψ describes microscopic processes while Σ reveals what we can hope to measure in regions of "space" where "diffraction" is unimportant. (The space is an abstract function space and not real space.) Diffraction is most easily comprehended in terms of an uncertainty relation. Such a relation may be derived from the commutator equation which is always associated with the Poisson bracket equation, (7). For simplicity let us assume we have one "particle", $N = 1$.

The commutation relation corresponding to (7) is

$$(\pi(\boldsymbol{x}'), S(\boldsymbol{x})) = -i\hbar\,\delta(\boldsymbol{x} - \boldsymbol{x}'). \tag{13}$$

If we average $\pi(\boldsymbol{x}')$ and $S(\boldsymbol{x})$ in (13) over two macroscopic spatial volumes $\Delta V'$ and ΔV, we find that (13) yields the uncertainty relation

$$\Delta \left| \int_{\Delta V'} \pi(\boldsymbol{x}')\, d^3 x' \right| \Delta \left| \int_{\Delta V} S(\boldsymbol{x})\, d^3 x \right| \simeq \hbar \Delta V\,, \tag{14}$$

provided there is a region of overlap between the two volumes.

We can also ask for the uncertainty relation (14) in the region of overlap, δx. If δx is small compared with the macroscopic volumes, the particle is most likely to be elsewhere in space, so that $\Delta(D\delta x) \simeq 1$. On the other hand, in the region of overlap

$$\Delta \left| \int S(\boldsymbol{x}')\, d^3 x' - \int S(\boldsymbol{x})\, d^3 x \right| = \Delta(dS/dx)\, \delta x\, \Delta V$$

and (14) becomes

$$\delta x\, \Delta(dS/dx) \simeq \hbar\,. \tag{15}$$

If we define the "particle" momentum as $p_x = dS/dx$ we have

$$\delta x\, \Delta p_x \cong \hbar\,. \tag{16}$$

Formulas (15) and (16) are indeterminacy relations that arise an attempted simultaneous measurements on the localization of a single "particle" and its phase. As we localize our "particle" its phase becomes more indeterminate. Formula (16) has the same form as the uncertainty relation (4) derived for an ensemble of rays. However (16) is significantly different since the uncertainty is related to a single "particle." True, the "particle" ($N=1$) is no simple object. It is highly non-local for the only precise information we have is that it is located somewhere in space. And only under proper conditions can we hope to see the "particle" manifest itself as a classical point particle. In general the "particle" can only be described by $\Psi(N=1)$, and it exists everywhere and "senses" all of space through its phase S. The microscopic particles which we observe in nature are eruptions of this extraordinarily complex "particle" which represents the unity of its identity (denumerability, $N=1, 2, \ldots$), and a "sensory" component, its phase.

The classical limit of the above sketched second quantized theory is the Hamilton-Jacobi theory, Eq. (12). The imagery of this classical theory is exactly the same as that described above for the "particle." The primitive classical images of point particle and wave (or phase) are limiting concepts of our more general "particle" picture, and they are most easily comprehended by means of the uncertainty relation (14). One should note however, that the primitive classical notions of particle and wave (phase) are closely tied to the Hamilton-Jacobi functional Σ, which is defined in terms of the constant N and the variable S. We know from classical dynamics that there are infinitely many other solutions Σ to Eq. (12) corresponding to other choices of constants and variables. These other "particles" will provide us with a richer classical imagery at the same level as the classical point particle and wave. The meaning and significance of these potentially new "complementary" classical concepts is still an open question.

In our presentation the phase variable S was not necessarily the phase of a wave. However, it is readily made so by requiring that it be a single-valued, continuous, and bounded function of position. If these requirements are made, the quantum theory of Equation (9) becomes the JWKB approximation to the second quantized Schrödinger equation. In fact for special choices of S and D one can construct classical entities, Ψ, which satisfy Eq. (9) rewritten as the exact second quantized Schrödinger equation.

However, quantization of (9) implies the imposition of specific restrictions on the functional Ψ. These restrictions in the function space are akin to those imposed in ordinary space on the wave functions of wave theories. These requirements in the function space give us the wave-particle duality, for they provide identity and phase for microscopic processes. The appearance of $\hbar$ in the first quantized Schrödinger equation is quite immaterial to these demands. From this point of view the predictions of the Schrödinger wave theory of distinct energy and angular momentum states have little to do with the quantum theory. These predictions are characteristic of all wave phenomena — the matter wave is only distinguished by its peculiar dispersion properties. In other words, wave functions of all wave theories (optical, acoustic, water, matter, etc.) depend on certain integers which arise because of the requirements of continuity, single-valuedness, and boundedness. The matter wave is slightly different from other waves in that the dispersion relation connecting wave velocity and wave length involves the strange constant $\hbar$. Real quantization, as distinguished from general wave properties, involves identity ($N=1, 2, \ldots$), and manifests itself in the space of the functional Ψ.

As is well known, the wave-particle duality is a general property of wave phenomena, e.g., it is exhibited by electromagnetic waves, elastic waves, etc. The mathematical procedure for predicting the duality is always the same and follows with slight variations the outline given above for matter waves. Not as well known is the fact that in all these theories, and just as for matter waves, one can find exact and approximate JWKB functional solutions to the counterparts of Eq. (9). Quantization is again achieved by requiring that the wave functional Ψ satisfy additional conditions. The precise requirements are related to the two types of quantum statistics. Since we are not prepared to discuss the general problem of quantum statistics from this point of view, let it suffice to say that in the second quantized Schrödinger theory both types of statistics are encompassed by our commutation relation, (13).

The concepts and the mathematical structure of classical theory and quantum theory are deeply intertwined. I feel that further study of the similarities and differences of these theories will lead to new physical insights regarding the properties of matter.

REFERENCES

[1] See P. G. BERGMANN, Basic theories of physics — mechanics and electrodynamics (New York: Prentice Hall, Inc. 1949), Eqs. (143) and (148), p. 258—259.

[2] BOHR, N.: J. Chem. Soc. 349 (1932).

[3] BARGMANN, V.: L. MICHEL, and V. L. TELEGDI: Phys. Rev. Letters **2**, 435 (1959). In the non-relativistic limit, and in a frame of reference where the electron is instantaneously at rest, the ordinary differential equations describing the spin motion are $\dot{s} = (e/mc)\, s \times B$.

[4] SCHRÖDINGER, E.: Brit. J. Phil. Sci. **3**, 109, 203 (1952).

[5] HEISENBERG, W.: The physical principles of the quantum theory, p. 10. New York: Dover Publ. Inc. 1930.

[6] BOHM, D.: Phys. Rev. **84**, 166 (1951), in which a similar variational principal is employed.

Chapter 10

Objectivity in Quantum Mechanics[1]

HENRY MARGENAU · JAMES L. PARK
Physics Department, Yale University
New Haven, Connecticut

1. The Meanings of Objectivity

1.1. Objectivity as Ontological Reality

Objectivity means many things. Some philosophers speak of an objective reality behind perceptible things, as that which in some sense *causes* appearances. They hold that the objective lies beyond human knowledge, beyond all experience but is responsible for it. When questions are raised as to the actual existence of an external world, independent of all knowers, it is *this* idea of objectivity that intrudes itself.

The first view to be considered, then, identifies objectivity with ontological existence, and it is a rather common view among physicists who are not given to philosophical reflection. The wholly unquestioning, of course, are satisfied with the attitude of naive realism which takes what is given in sensation to be objective and real. At this stage, the ideas of objectivity and reality are fused together; the refined considerations which force a separation between them have not arisen.

But not many scientists, let alone quantum physicists, are *naive* realists. For if one seeks the objective, understood as the cause of sensations, in the things that appear in sensation, one's search is at once led beyond appearances, since even the simplest scientific observations show that things are not as they are perceived. Atoms and molecules cannot be perceived directly; and if they are to be regarded as "things" in the same ontological meaning as the things we see, a change in the connotation of that word is required. This is particularly necessary when it is realized that the constituents of the atom, being smaller than a wave length of light, cannot be carriers of color; being subject to the uncertainty principle they cannot always possess determinate positions or sizes; in short, when it is realized that they may not be endowed with sensory qualities at all.

[1] Work supported by the United States Air Force Office of Scientific Research.

Moreover, even realism made sophisticated by the admission that objectivity does not lie within sensation but resides in transcendental entities like atoms, elementary particles, etc., which occasion sensations, is haunted by the fact that the point of contact between the objective and the sensations it produces, i.e., the act in which the real reveals itself to the knower, is shrouded in subjective mystery. This poses a dilemma which troubles all who equate the objective with the ontologically real: The objective is to be inferred from sensations as their common cause; yet it is doubtful whether sensations have anything in common at all. Nobody can show that the color I see is qualitatively the same sensation as the color seen by you. For example, sensations produced by the same objective state of affairs do differ in people who are colorblind.

These are the obvious difficulties of the view in question. If they can be removed, perhaps by postulating that there is a common element in human perception of the world, objective realism has much to recommend it. For it allows itself to be coupled with an old physiological theory of perception which has a solid background in science, leading to an account such as this.

The objective world, which is beyond all experience, is composed of those entities which the scientist, primarily the physicist, continues to discover. They have objectively real attributes like mass, charge, size, shape, position and velocity. They cause objectively real effects in the form of light and sound which our sense organs can receive. The stimulus then travels, again as a physical impulse, to certain places in our brain where translation into conscious response occurs, and a sensation arises.

Clarity and simplicity favor this theory; wide acceptance almost saves it from criticism. It can even explain why sensations differ. To do this it need merely invoke differences in the physical make-up of percipients. Yet there are two major problems which it cannot resolve; one is philosophical, the other physical.

The philosophical problem is to account for the conversion of the physical stimulus into a conscious response, i.e., the age-old mind-body problem. Its solution requires such extraneous devices as the doctrine of psycho-physical parallelism or the Marxist view that consciousness is a manifestation of matter at a certain level of complexity. At any rate, the passage from physical stimulus to conscious response cannot be regarded as an ordinary example of a cause producing an effect; for these, as commonly understood, always act within conscious experience and cannot link the non-conscious with the conscious, nor that which is outside experience with an item of experience.

The physical problem raised by the view in question has its roots in quantum mechanics. Classical physics permitted us to endow the ultimates of the atomic world with those qualities which carry the accent of objectivity in the world of ordinary experience, the qualities on which all reasonable men agree, e.g., position in space, speed, self-identity, size, and mass. These, however, are exactly the attributes whose assignability to the ultimate constituents of the world quantum mechanics has taught us to doubt. As we shall see later, these attributes are closely related to the measurement act, may indeed be engendered by observation, by perception. And thus one is in danger of affirming that the objective aspects of reality beyond experience are somehow dependent on the subjective choice of what one decides to perceive.

A final difficulty in this thesis which identifies objectivity with ontological reality arises in the historical fact that conceptions of the supposedly real entities inferred from sensation change in time. Yet surely the ontologically real should exhibit a high degree of immutability. At this point, ontological realism may cling to a very general concept, like matter, define it in a rather indefinite way which leaves room for changes, and pronounce it the quintessence of the objective. For a time this may succeed, but there are already indications which render the posture of dialectical materialism unreasonable. They are present in the recognition that very unmaterial species of onta, matterless particles and fields, play essential roles in quantum mechanical theories; it may even be that probability fields are irreducible constituents of the ontologically objective. These leave the concept of matter far behind. Then, if we wish to continue to play the game, we have to face these esoteric features and pronounce them objective — for they are after all the residue of what began as an objective thing. And in doing this we could not be sure that fifty years from now an entirely different objective picture would not confront the scientist.

Clearly, what is needed to make an approach of this sort attractive is an act of faith, a postulate of a non-ontological sort which expresses the conviction that the progress of science converges upon a final limit of "truth". This is indeed the conviction which inspires research in science. On the basis of it one can formally define reality to be the goal of the scientific enterprise. Objectivity then is not given but posed as a problem. (One recalls here NATROP's phrase, Das Wirkliche ist nicht gegeben, es ist aufgegeben.) The kind of objective reality thus designated might be called asymptotic.

About it two things must be said. First, it resides no longer beyond experience in a manner that makes it wholly different from experience; for it stands at the end of all experience. It is ideal inasmuch as it is never within human grasp. Objectivity thus becomes an ideal and cannot be assigned to any present phase of scientific operation. Worse, however, is the fact that we do not know it and cannot talk about it in meaningful terms. We therefore dismiss ontological objectivity from further consideration in this article, since it would not allow us to answer the question whether quantum mechanics, a theory currently known, involves or does not involve objective elements.

At this stage, then, we turn to definitions of the word objective that seek its substance within scientific experience, not beyond or at the invisible end of it.

1.2. Objectivity as Intersubjectivity

Since we are now beginning to rely heavily on an understanding of the word *experience,* it is well to state clearly what is to be meant by it. Its root is, of course, the Latin *experiri,* which has a very wide connotation including feeling, sensing, thinking, indeed practically all modes of awareness. Because of its catholicity of meaning it defies clear logical definition. This should not, however, be taken as an indictment, for very few matters that concern us strongly are capable of explicit definition.

The other meaning of experience, which has come to dominate modern thinking, stems — by strange substitution — from the Greek word "en peira ('in trial')". It denotes outer experience, the contingent perceptions and the data that assail us from without. This meaning equates experienced with empirical. Now it will be evident to everyone conversant with the quantum theory that it cannot get along with only that part of Latin *experiri* which the Greek *en peira* singles out. It must include at least the concepts and relations in terms of which science explains its data, and they are not empirical in the narrow sense. The contraction of the pristine meaning of the word experience occurred as the result of the movement known as British empiricism. To do justice to quantum mechanics, it is necessary to restore its original significance and denote by it *all* phases of awareness. For our purposes, however, we may lose sight of affective and conative experiences and focus attention on those which lead to knowledge as distinct from feelings.

Strictly, all experience is first-person experience, and this is almost by definition and at least in the common understanding of the term, *subjective.* But subjective is presumably the opposite of objective. The problem, then, is how to remove the subjective element from first-person experience, that is to say, how to eliminate those features which originate in the person having the experience. Perhaps this cannot be done; perhaps the very act of perception involves, as KANT believed, ingredients contributed in universal ways by every percipient. If this were true communality of perception among many subjects would not remove the personal, or the human, admixture from experience. And if that admixture represents the subjective, it remains immune against the remedy of communality. The Kantian, therefore, would be unimpressed by the view we are now discussing, the view according to which *objectivity* is *intersubjectivity*. To others the thesis has great appeal, and it is foremost in popularity among the views here surveyed.

According to it individual experience cannot be trusted. Everybody knows about sensory illusions, vivid dreams and hallucinations. Only what is trustworthy is worthy of being called objective. Thus, to single out the objective, one may use a method designed to establish "truth". If witnesses of an occurrence differ in their accounts, common sense seems to compel us to dismiss contradictions as untrue and to retain as anchored in truth those parts of their reports which are common. We need not discuss here the philosophical reasons for this compulsion, nor the justification of the procedure. What matters is that the present criterion of objectivity, i.e., intersubjectivity, adopts a common procedure for establishing the truth of reports for the purposes of guaranteeing objectivity of experience. The basic assumption is that the objective content of one first-person experience coincides with that which others report as having experienced under similar external circumstances.

This method, then, leads to the discernment of objectivity in the world of sensations. But the intersubjectivity argument can likewise be used in the realm of ideas. External happenings, common observations suggest conjectures and speculations with respect to unobservable entities; and, in this passage from facts to ideas, intersubjective report of observations is neither a reliable nor a coercive guide. There are many possible interpretations of objective events in scientific theory; and it is becoming increasingly clear that the so-called method of induction, which has at times been supposed to lead with cogency from facts to theory, is quite inadequate to assure the objectivity of theoretical constructs as abstract as those of quantum mechanics, even when the data satisfy the requirements of intersubjectivity. Hence this requirement needs to be applied to ideas also and separately: scientific theories are correct when scientists agree upon them. In this way, theories and the ideas they convey become objective.

Perhaps it is well to make a distinction between two groups who hold views like that under discussion: α) those who use intersubjectivity only for discriminating between subjective and objective *sensations* and either regard ideas as subjective or invoke other criteria for establishing their objectivity, and β) those who rely upon agreement to assure objectivity in both spheres. Evidently class β is more vulnerable than class α because it is harder to agree upon ideas than upon facts.

Specifically, the difficulty with the intersubjective acceptance of interpretations lies in the assignment of competence, in the weighting of the judgment of those who agree or disagree. Only those who understand a theory can rightly be included among contenders, and their number is surely indefinite. If EINSTEIN had counted noses among those who agreed and disagreed with his conclusions regarding the contraction of moving objects or the renowned twin paradox, the objectivity of his theory, judged on these grounds, would have been extremely low. And it might not be high today in view of all the amateurs and half-baked logicians who claim that his conclusions contradict reason. Again, if the writings of all philosophers who deal with the uncertainty principle are placed into the balance, the physicist's ideas may well be thought subjective.

Another difficulty for the class β protagonist arises from the fact that a theory which he must now regard as objective may lose this quality in time. What is troublesome here is not the fact of change — objectivity as well as truth may change in time — but rather the circumstance that the criterion of intersubjectivity provides no reason at all for understanding why a change occurs. If people alter their views there must be reasons aside from the desire to be in the majority which induce at least some of them to do so. Hence it is clear that intersubjective agreement on ideas cannot play the full role of defining objectivity among theories.

While the case is a little stronger for those in class α, even here troubles appear. There are reports of mass hallucinations; large audiences have been subjected to sense deceptions; magicians feel more secure when performing tricks before crowds than in the company of few. These objections, however, seem trivial when compared with the realization that the criterion in question is often irrelevant: One can, without reference to other people, convince himself that a certain observation was erroneous, i.e., non-objective. One of the authors, for example, on one occasion when occupying an office in the Physics Laboratory of the University of Washington, saw a beautiful white cloud hovering some distance from his window. When he commented on this, nobody contradicted him. But the cloud stayed put for days. In his amazement he looked at the map and discovered, all by himself, that he had been gazing at the snow cap of objective Mt. Rainier. It seems as if objectivity can sometimes be distilled in subtle ways out of the experience of a single person.

This possibility gains further interest when it is realized that discovery, the bursting of objective truth into a single mind, is rarely a collective phenomenon. Surely, whenever possible, others will repeat the experiment that led to the discovery and agree with it. Yet there are instances, like the observation of the birth of a nova, which by their very nature are unique. Here the scientist takes the word of a solitary astronomer tentatively to be true and checks it *against other knowledge*, often first-person knowledge. The criterion of validity here is not intersubjectivity but *theoretical consistency* of a certain kind.

Noting this, we now examine theories of objectivity which, while welcoming whatever guarantees the property of communality can offer, seek objectivity within the context of one person's experience.

1.3. Objectivity as Invariance of Aspect

When a thing is seen, many of its properties depend on the relation between the viewer and the object, on his distance, his perspective, the angle from which he sees the object. Shape is one of these properties, e.g., a circle is an ellipse when seen from an angle. Color, size and position with respect to nearby things are others. Indeed it is hard to find observable properties which are not relative and therefore subjective in this sense. Stability does seem to go with a few attributes, like weight and volume, and in others, situations can easily be created in which invariant aspects reveal themslves. For instance, a circle will never look like an ellipse on frontal view, color will not change in the same specified illumination, size of one object will be invariant when it is observed from a standard distance.

The theory at issue holds that objectivity must be assigned to those properties which are, or can be made, invariant. Something is objectively round if under the same specified conditions it always appears round; it is objectively blue if it appears blue in sunlight, and so on. More needs to be said about this theory when it is applied to ideas and interpretations. For the present we deal again with aspects of immediate experience, where this criterion suffices, and again we call this thesis α.

We have already noted that practically nothing is invariant except under specified conditions. Not even weight and volume, which we singled out as nearly objective without conditions, exhibit this character fully. For they are not invariant when crudely apprehended; e.g., when bodies are weighed or spanned by hand. *Balances* make weight, *volumetric procedures* make volumes invariant. Indeed, all sensations, all direct outer experiences, lack the stability in question unless they are severely restricted in carefully prescribed ways. What we are saying is that invariance, and hence objectivity, is conferred by instrumental procedures. We encounter here the problem of operational definition.

The "temperature" I feel in my fingertip when I place it in a hot bath is a subjective sensation in every sense of that word, since it leads to differing reports from different individuals depending on the recent history of the fingertip. It even changes "subjectively" in my own experience when I transfer my finger from an ice bath to hot water. Invariance is achieved, however, if my finger is replaced by a thermometer; the *measured* temperature is invariant and therefore in this sense objective. The procedure in question is called by philosophers, with some pretense to erudition, an operational definition of temperature. Physicists call it simply a measurement.

Every measurement employs an instrument, and in many cases the measuring device seems to be an extension of, or a refinement upon, our sense organs. The use of instruments for the sake of producing objective knowledge is therefore often regarded as philosophically trivial, since they merely enhance the normal method of acquiring factual knowledge. This attitude, however, is fallacious. A measurement does more than improve the accuracy of our normal senses. It establishes a new item in experience, an item somehow correlated with the subjective quality it expresses but not identical with it.

The temperature measured by means of a thermometer, i.e., the point of coincidence between the top of a mercury column and a scale, is different in every sense from the sensation in my fingertip; the weight registered by a balance differs totally from the weight sensation of holding an object in my hand. The force recorded by a dynamometer is not the kinesthetic awareness of a push or a pull.

We conclude, therefore, first, that according to the invariance criterion no immediate sensation is objective; second, that science uses the procedures of measurement to set up invariant counterparts to variant and hence subjective immediate experiences. Strictly speaking these counterparts are contrived, constructed vis-à-vis the flux of sensations; measurement provides rules of correspondence between constructed invariances and the items of direct perception. According to the present theory (part α) measured quantities and the entities, bodies, or systems to which they refer, can be said to be objective attributes or parts of the universe.

The going is rougher when we examine part β of the invariance thesis, which assumes that *ideas* attain objectivity through invariance. This might mean that a certain concept is encountered as the result of manifold avenues of reasoning. In quantum mechanics the question is often asked whether the probabilities assigned to observable events are objective or not. From the present point of view the answer is affirmative if different considerations, indeed all relevant considerations, lead to the same idea and the same value of the probability. Such, we feel, is the case in quantum mechanics. Note however that this kind of objectivity does not satisfy the ontologist, whose attitude we discussed in section 1.1 and who wants to know whether in some transcendental sense events are free from probabilities, since they either happen or not.

On the other hand, the invariance view, conceived as stability of an inference against all proper modes of reasoning, rules out certain conjectures like the wave-nature of electrons because it is the consequence of one set of arguments and not of another.

The technical meaning of invariance in recent physics, while most important in characterizing theories that are likely to be successful in a general sense, is too specific to have much relevance for the problem of objectivity. It is a property of certain mathematical descriptions of natural processes and leads to relativity in the phenomena described. Thus, according to the special theory of relativity, the four-dimensional metric is invariant with respect to certain transformations, while spatial and temporal relations are not. If this were to be interpreted by saying that the metric is objective but spatial and temporal occurrences are not, few physicists would agree.

Similarly, objectivity in quantum theory has sometimes been identified as invariance relative to complementarity[1]. L. ROSENFELD [*1*] has elaborated this viewpoint as follows: "Physical quantities ... correspond with operators susceptible to an infinity of numerical representations. Each of these representations refers to particular conditions of observation, but the equations connecting the operators are invariant for the canonical transformations which express the passage from one mode of observation to another. These equations represent the objective content of the theory, the objective expression for the quantal laws of nature."

In reply to ROSENFELD, MARIO BUNGE [*2*] has suggested that the terms "absolute" and "objective" are erroneously used as synonyms. We note here, in agreement with BUNGE, that the Rosenfeld objectivity inheres in a rather technical invariance which may lose contact with the central meaning of that concept. In the realm of observations the criterion of invariance serves a useful function. And it makes sense to a certain extent in the world of theory although one can hardly subdue the feeling that invariance of theoretical aspect *alone* is not decisive.

A more tenable version of the thesis that in quantum theory objectivity appears as invariance relative to complementarity has been propounded by MAX BORN [*3*]: "I think the idea of invariant is the clue to a rational concept of reality ...

"The final result of complementary experiments is a set of invariants, characteristic of the entity. The main invariants are called charge, mass (or rather: rest-mass), spin, etc.; and in every instance, when we are able to determine these quantities, we decide we have to do with a definite particle. I maintain that we are justified in regarding these particles as real in a sense not essentially different from the usual meaning of the word."

---

[1] A concise statement of BOHR's famous principle is the following: "Evidence obtained under different experimental conditions cannot be comprehended within a single picture, but must be regarded as *complementary* in the sense that only the totality of the phenomena exhausts the possible information about the objects." NIELS BOHR, *Albert Einstein: Philosopher-Scientist*, p. 210.

1.4. Objectivity as Scientific Verifiability

Science has evolved a method for determining what it regards as *acceptable judgments* in the face of the evidence available at a given time. These judgments reveal its commitments with respect to truth and reality; under the present heading we equate objectivity with scientific truth.

This truth cannot be absolute because it obviously changes as new evidence appears. Whether physicists fifty years from now still believe in electrons as elementary particles is highly questionable; yet according to the view now under discussion electrons are today part of the objective world. If objectivity means permanence of conception, only the ontological interpretation (1.1) is tolerable, and we have seen how its vagueness, its lack of verifiability, make it unsuitable for scientific use.

The method of verification, too, is not fixed in the nature of things, nor is it unalterably grounded in the mind or brain of the knower. Its components are being written while science is made; success of explanation, correct prediction, survival against the vicissitudes of change are its guidelines, and its product is objectivity. The method is immanent, not transcendental; it makes no appeal to ontological reality yet is never forced to disavow it. Operating wholly within experience it relies upon certain organizing principles exhibited by that experience to define the objective.

These principles were studied in a previous publication [*4*] and will here be only briefly reviewed. First, a distinction is made between non-cognitive and cognitive experience, and objectivity is placed within the latter. In it are recognized certain components, each with a character of its own. These are the datal protocol experiences, of which a sensation is typical, displaying a high degree of contingency and coerciveness. Near the opposite pole are ideas, concepts, with the chief peculiarity of having been *constructed* by the experiencer. Protocols are always subjective in the beginning — they are the seen color, the felt temperature, the sensed pull called a force. But by the device of operational definition, i.e., by measurement and more generally by certain rules of correspondence, the subjective protocol experiences are made objective in ways that have already been discussed in 1.3.

Strictly speaking, measurement converts raw data into constructs, items of experience for which man himself is responsible. All observables whose symbols appear in equations are constructs — not sensed qualities — in this understanding; and they are objective by virtue of being invariant. Barring minor discrepancies which are dealt with in the theory of errors, everybody measures the same temperature with a given thermometer, perhaps in contradiction to the verdict of his fingertip. But now we are facing the question whether this degree of invariance, conveyed by the standardizing procedures of measurement, is enough to insure objectivity of all theoretical entities, of all constructs.

Operational definitions are arbitrary. One can measure temperature by a variety of different thermometers and obtain different values, each invariant with respect to its own instrument. Thus arises the question: which is the objective temperature, that registered by an alcohol thermometer, that recorded by a mercury thermometer, or the one read from the ideal gas scale? If science were merely a discipline for making accurate measurements one might call them all objective. But temperature, and all measurable quantities, are meant to be significant of something in a deeper sense. Temperature bears reference presumably to the speed of molecules, and objectivity should also apply to them. Their objectivity, however, can hardly be established by invariance of measurement procedures.

On the other hand, it is clear that certain operational definitions, for instance that based on the ideal gas or the Kelvin scale, have special qualities to recommend them. If, for example, I affirm that temperature is proportional to the mean kinetic energy of the molecules in a body, I must prefer KELVIN's operational definition to that involving an alcohol thermometer. An important fact comes into view at this point, the fact that theory, scientific law, can often discriminate between measurement procedures and show that some are good and others bad. And this judicial function of theory, according to the thesis under study, must be respected in defining objectivity.

With that understanding we continue our survey of the verifying method. The passage from sensation to quantitative constructs makes the data stable, as we have seen. But it does more; it affords the possibility of reasoning about them. There is not much one can do by way of logic or mathematics about the temperature sensation in one's fingertip, but a great deal about the measured number 90° F. Hence the "rules of correspondence" have a dual purpose, to stabilize and to rationalize. Among the constructs we can reason. But in the beginning they are freely chosen, and the latitude of choice is so great that science would flounder if all constructs that are operationally definable or otherwise stand in correspondence with protocol experiences were equally admissible. Hence there must be regulative and discriminative principles which limit that choice.

A large literature is devoted to these regulating principles. They are vaguely referred to by the metaphor of OCCAM's razor, by economy of thought, and occasionally (though erroneously) by the inductive method. In reference [*4*] they are outlined as a spectrum of "metaphysical" requirements and discussed under the names of logical fertility, multiple connections, stability, extensibility, causality, simplicity and elegance.

None of these requirements can be satisfied absolutely; each makes its demands in competition with the others, and the seasoned scientist knows somehow when maximal justice has been done to them all.

The metaphysical principles are not alone in deciding whether a conjecture is verified because they do not provide the solid links with observation, with protocol experience on which certified knowledge depends. Hence they are augmented by well known processes of empirical confirmation. Perhaps the word "augmented" is too weak to express the crucial importance of the confirming act, and most scientists would insist on placing it before the regulative principles as the means for verification. The point we make is that both are needed, that one is complementary to the other.

Constructs which satisfy the metaphysical requirements as well as the stringent rules of empirical confirmation are called verifacts, and verifacts are the carriers of objectivity in the domain of theory.

Admittedly, this definition does some violence to the common-sense implications of objectivity, as does almost every other careful definition. It suggests, for example, that concepts like the state of a particle which is given by a wave function, though not directly accessible by protocol experience[1], can nevertheless be objective; that as was already mentioned, objectivity of an entity may cease in time; that interesting mathematical constructs which do not refer to the world of observation lack objectivity. The definition contradicts in particular the literal allusion of the word, which seems to refer to an object. If by object is meant a thing, our version of objectivity is much too generous, for it includes innumerable concepts that are non-material and abstract.

It is our belief that the accounts labelled 1.3 and 1.4, invariance and verifiability, are the most defensible on philosophic grounds and at the same time closest to the understanding of objectivity among physicists who work in the quantum theory.

1.5. Intersubjective Subjectivity in Science

The title of this section is not a contradiction in terms. It has already been explained in section 1.2 that, while communality is invaluable to science, it is by no means sufficient to guarantee objectivity. Moreover, although the physicists whose ideas are about to be presented do speak of subjectivity in science, none would deny the simultaneous presence of intersubjectivity.

An examination of philosophic views which attribute to the quantum theory a subjective aspect reveals that the term *subjectivity* is used ambiguously — sometimes in a single paper. The two dominant meanings can be distinguished by using the qualifiers probabilistic and Kantian.

[1] Although its occurrence can be confirmed by measurements.

Probabilistic subjectivity is a concept applicable only to theories which involve probability and statistics. It is rooted in the doctrine that probabilities represent degrees of knowledge. The remaining sections of this article are devoted to the question as to whether quantum mechanics exhibits subjectivity in this sense.

Kantian subjectivity, on the other hand, is a much broader notion, central to the understanding of the scientific method in general. It is the "subjectivity" apparent in statements such as HEISENBERG's assertion that what is observed by scientists is "not nature in itself but nature exposed to our method of questioning" [5]. The words *idealistic* and *a priori* better convey the intended meaning of *subjective* in this Kantian sense. For example, in the terminology of the preceding section, verifacts might be labelled subjective in order to emphasize their constructional aspect; but such a label runs the risk of committing unintentional distortion. Nevertheless, the term *subjective* is sometimes used in the Kantian sense; and this usage entails certain ideas of crucial significance for the philosophy of science, in recognition of which we advert briefly to the writings of BORN and EDDINGTON.

BORN notes, in a recent essay [6], that "fundamentally everything is subjective — everything without exception", thereby emphasizing the first-person character of experience (section 1.2). Intersubjectivity is then established but this in itself does not confer objectivity. In his search for objective knowledge within subjective experience, BORN is led to the mathematical constructs in the exact sciences.

"Mathematics", he says, "is just the detection and investigation of structures of thinking which lie hidden in the mathematical symbols ... These are structures of pure thinking." He concludes that in theoretical physics "hidden structures are coordinated to phenomena; these very structures are regarded by the physicist as the objective reality lying behind the subjective phenomena."

In the language of preceding sections, BORN is suggesting that the verifacts of 1.4 transcend their experiential realm and indeed describe the ontological reality of 1.1. The structures in scientific theory are, for BORN, identifiable with the Kantian *Ding an sich*.

Finally, the "selective subjectivism" [7] of Sir ARTHUR EDDINGTON seems to exemplify the notion of communal subjectivity. This is not the place to attempt a survey of EDDINGTON's fascinating scientific epistemology. Suffice it to say that he lays great stress on the *constructional* nature of verifacts. Even though he favors the idea of an objective ontological reality, EDDINGTON, unlike BORN, does not regard the verifacts of science as identifiable with transcendent components of that reality. Perhaps a succinct expression of his central thesis is in this final colorful passage from one of his books [8]: "[Scientific knowledge] is knowledge of structural form, and not knowledge of content. All through the physical world runs that unknown content, which must surely be the stuff of our consciousness. ... And, moreover, we have found that where science has progressed the farthest, the mind has but regained from nature that which the mind has put into nature.

"We have found a strange foot-print on the shores of the unknown. We have devised profound theories, one after another, to account for its origin. At last, we have succeeded in reconstructing the creature that made the foot-print. And Lo! it is our own."

2. Probability

Like many recent scientific theories, quantum mechanics operates extensively with probabilities. In the eyes of many, this alone makes it suspect, for are not all probabilities subjective? To answer this question, we review the idea of probability.

It had a humble beginning in men's concern with games of chance. In its first appearance on the mathematical scene it was taken as an index of confidence in the outcome of an event. Quantified as odds, it expressed a person's expectation of some future happening. Confidence and expectation are private matters which may well vary from person to person. Hence it is easily concluded that probabilities are always subjective estimates of likelihood. This result is in keeping with some modern theories of probability, which are sometimes called a priori theories and occasionally even subjective[1]. Their essential starting point is an insight of LAPLACE who defined probability as the ratio of the number of favorable events to the total number of "equipossible" events. To illustrate, the probability of throwing a five with an honest die is $\frac{1}{6}$ because the die has one face marked with a five and six faces altogether. But evidently, this ratio has nothing whatever to do with the outcome of the next throw, or with the outcome of any one throw of the die. For some reason, not wholly clear from this definition, the ratio is a measure of the confidence one ought to have in the occurrence of a five, and one seems justified in using it as a guide in betting. As an index descriptive of the die the ratio is indeed objective, but as a probability it is not; it represents a subjective measure of likelihood insofar as it refers to an actual event or, put differently, a degree of knowledge concerning that event.

If this interpretation is maintained consistently, the probability must change when the event occurs. Thus, when the die is thrown and a

[1] To quote H. JEFFREYS, *Theory of Probability*, Oxford Press, (1939): "In fact, no 'objective' definition of probability in terms of actual or possible observations, or possible properties of the world, is admissible."

five appears, the probability has changed from $\frac{1}{6}$ to one; in any other outcome it jumps to zero. The ratio, to be sure, has remained the same, but the facts entailed by the probability interpretation have belied its pretension.

Many difficulties beset this subjective view; among the most troublesome is our inability to specify in many instances the number of equipossible events which enter the ratio. What is the probability that an unknown person is a thief? It is natural to conclude that the probability of his being honest is $\frac{1}{2}$, which is clearly absurd. This is not the occasion to comment on the numerous rescue efforts that have been made to save a priori probabilities and their subjective implications. Suffice it to say that most sciences, especially quantum mechanics, take an approach which avoids this kind of subjectivity.

In their version, probability is a relative frequency of events, or the mathematical limit of relative frequencies. To find the probability of a five one throws the die n times, counts the number of times, say n_5, a five appears, and forms the ratio n_5/n. For small n this ratio fluctuates, but as n increases the fluctuations becomes small and rare. Probability thus defined has reference not merely to the die but also to the sequence of throws; it does not change significantly when a further throw is made — subjective jumps to the value 1 or 0 do not take place. In the sense of 1.3 and 1.4 therefore this probability is an objective quality of a series of throws. The price one pays for invariance is the surrender of meaning with respect to single events; for the frequency definition necessarily involves an aggregate of cases and becomes powerless when confronting a unique occurrence. It cannot handle such notions as the probability that the universe shall cease to exist tomorrow since there is no series of observations of which this is an instance, nor can a single event ever serve to determine a relative frequency.

So far, then, it seems as if there were two probability ideas, one partly subjective and one objective, and that only the latter has a place in science. Closer study however reveals an interesting connection between the two. It is, after all, the case that the two *agree numerically*, although they are logically unrelated. To state this seemingly miraculous coincidence is to invoke a law of nature, a proposition which equates a theoretical construct, the a priori ratio of faces, to a measurable quantity, the relative frequency. The logic of the situation is analogous to that which surrounds every theoretical law. Consider, for example, the law of gravitation: the product of the masses divided by the square of a distance is a construct which is logically unrelated to the operational meaning of a force; yet the law asserts that they are equal and measurement bears this out. We have seen earlier that every fully formulated scientific quantity must have a dual reference, once to datal observations and once to theoretical constructs. Probabilities as they are used in science satisfy this rule. They can be measured, objectively determined, by means of the frequency definition; they can be theoretically predicted by — in the simple instance of the die — LAPLACE'S formula.

In general, the theoretical formula for calculating probabilities is more complex than this, even in games of chance. Basically, that formula suffices for computing the chance that three aces will be dealt to one bridge player, although it needs to be used with care. It does not serve, however, to suggest the probability that two electrons in a hydrogen molecule will be found attached to the same atom. A different a priori definition is required here. The squared modulus of a ψ-function is a probability, perhaps subjective if it is interpreted as an expectation of what might be found in a single observation. It yields a formula for computing and predicting, just as did LAPLACE'S ratio. But when $|\psi|^2$ is coupled, through the laws of quantum mechanics, with the outcome of a series of observations on the position of the "particle" represented by ψ, the subjectivity disappears, the concept $|\psi|^2$ and relative frequency merge into a single meaning which is objective according to our accounts in 1.2, 1.3 and 1.4.

Many philosophers and a fair number of physicists evince displeasure at the thought that probabilities should be accepted on a par with measurable physical quantities, like lengths and sizes and masses. They feel that there is a difference which relegates probability to an inferior status, makes it imprecise and untrustworthy. It is therefore said that the use of probabilities must be indicative of incomplete analysis, a sign that something crucial has escaped detection. This is true in certain situations where probabilities are used for scientific convenience, but it is not conditioned by the nature of probability as a scientific concept. As such it is just as clean, secure, and determinate as any other scientific quantity.

The disturbing feature which seems to contradict this remark is the fact that a probability cannot be determined, measured, in a single act. The number n must be large to make the frequency definition applicable. On the face of it, this contrasts with the measurement of an ordinary quantity, like length, whose value can be read once from a scale. The fallacy of this reasoning need hardly be emphasized here, for every experimenter knows about the vagary of observations. If a good measurement is repeated, a different value will emerge practically every time; and a long series of readings will always result in a distribution of values from which, by statistical consideration, a "true" value can be inferred. To achieve this, the theory of probability enters. Thus, to put it bluntly, the concept of probability is even prior to that of a quantity like length. A single measurement fixes neither an ordinary physical quantity nor a probability, and there remains no reason whatever for excluding the latter from the class of ordinary, decent, measurable physical attributes of the world. Nor can it be denied the quality of objectivity, provided it is taken in its full theoretical-empirical context.

To exemplify antipodal interpretations of quantum mechanical probabilities, we now present typical quotations from the sides of subjectivism and objectivism.

Sir JAMES JEANS, famous for his highly subjective characterization of Schrödinger waves as "waves of knowledge", has written [*9*]: "The wave-picture does not show the future following inexorably from the present, but the imperfections of our future knowledge following inexorably from the imperfections of our present knowledge." In fairness, it should be noted that Jeans means communal, not personal, knowledge [*10*], and therefore confers on quantum theory that degree of objectivity discussed in section 1.2.

In spite of the empirical success of quantum theory, physicists of such stature as EINSTEIN and SCHRÖDINGER have pronounced the theory incomplete on the ground that physics is properly concerned with *objective* physical reality. According to K. R. POPPER, their requirement can be met without eliminating probabilities from quantum mechanics, provided probabilities are understood not as subjective measures of knowledge but in the light of that author's propensity interpretation.

This interpretation is fundamentally in accord with our discussion of the relation between a priori probability and relative frequency. POPPER [*11*] takes "as fundamental *the probability of the result of a single experiment,* with respect to this *conditions,* rather than the frequency of results in a sequence of experiments," although an experimental sequence is undeniably required to test a probability statement. "But now the probability statement is not a statement *about* this sequence; it is a statement about certain properties of the experimental conditions, of the experimental set-up." Probabilities are thus said to "*characterize the disposition, or the propensity,* of the experimental arrangement" to yield certain relative frequencies in ensembles.

POPPER believes that his propensity interpretation "takes the mystery out of quantum theory, while leaving probability and indeterminism in it." SCHRÖDINGER'S ψ-function is said to determine "the propensities of the states of the electron". If these propensities are regarded as objective attributes of the electron, even though they are measured statistically, the ψ-function may indeed reasonably be considered descriptive of objective, physical reality. This view does not differ in its philosophic content from the one put forth by one of the present authors [*12*].

3. The Crucial Issues in Quantum Theory

3.1. Von NEUMANN's Theory of Mixtures

In quantum mechanics objectivity hinges upon the deeper interpretation of the statistical elements of the theory. To resolve the problem of objectivity in quantum mechanics, it is therefore necessary to understand the mathematical description and classification of quantum statistical ensembles.

The statistics of such an ensemble are characterized by a statistical operator ϱ (whose representation is called the density matrix), which is related to observations through the "mean value postulate": If A is the operator corresponding to some observable (also called A), then the expectation value of A is given by the formula

$$\bar{A} = \mathrm{Tr}(\varrho A). \tag{1}$$

VON NEUMANN [13] has shown that the axiomatic correlation of observables to operators, together with general principles of statistics, permits the rigorous classification of quantum mechanical ensembles into two types: pure and mixed. By definition, an ensemble is pure if no subdivision thereof into two subensembles with *different* statistical operators is possible; a mixed ensemble is simply one that is not pure.

In terms of ϱ, the necessary and sufficient condition for a pure ensemble is that

$$\varrho = P_\psi$$

where P_ψ is the projection operator into the closed linear span of the vector ψ. The state of a pure ensemble is thus completely described by a single vector, the famous state vector ψ of quantum mechanics. Accordingly, for a pure case, equation (1) becomes [for $(\psi, \psi) = 1$]

$$\bar{A} = \mathrm{Tr}(P_\psi A) = (\psi, A\psi), \tag{2}$$

the usual statement of the mean value postulate in the quantum theory of pure cases. Equation (2) can be inverted to give the probability w_n that a measurement of A yields the eigenvalue a_n,

$$w_n = |(\psi_n, \psi)|^2, \tag{3}$$

where ψ_n denotes the eigenstate belonging to a_n. (We are assuming eigenvalues to be discrete and non-degenerate throughout this discussion.) Of special importance here is the fact these probabilities w_n, since they are associated with *pure* ensembles, represent maximal precision in quantum theory; they are therefore called irreducible probabilities.

The mixed ensemble, on the other hand, by its very definition, admits of subdivision into component pure subensembles, and will therefore require for its description not only irreducible probabilities but also probabilities analogous to those in classical physics. For definiteness, consider the mixed statistical operator

$$\varrho = \sum_n W_n P_{\varphi n},$$

The mean value postulate gives

$$\bar{A} = \mathrm{Tr}\left(\sum_n W_n P_{\varphi n} A\right) = \sum_n W_n (\varphi_n, A\, \varphi_n). \tag{4}$$

Equation (4) suggests the interpretation that a mixture state is to be regarded as a set of pure states φ_n with respective weights W_n. Thus the mixture state involves just that measure of ignorance which requires the use of probabilities in classical physics.

In statistical mechanics, for instance, one supposes that each molecule is in a definite dynamic state (q_i, p_i), only one does not know nor does it matter which particular molecule occupies that state. It is therefore necessary and proper to introduce the relative number of molecules, W_i, which partake of the state (q_i, p_i). On the basis of this mixture of knowledge and ignorance all theorems of statistical mechanics can be established.

In quantum mechanics, as we have seen, ψ takes the place of the classical state (q_i, p_i). A mixture assigns to every ψ a relative frequency of occurrence W_i, and if all W_i are known the theorems of quantum statistics can be proved.

Now the W_i have a special property: they are *reducible* probabilities That is to say, it is possible by selection, refinement of observation, or some other physical contrivance, to change these probabilities from whatever value they have to 1 or 0. As explained above, this is not possible for the *irreducible* probabilities that inhere in ψ.

In what sense are these probabilities to be conceived? The answer to this question holds the clue to the problem of objectivity in quantum mechanics.

Involved here is the philosophical and to some extent mathematical interpretation of the measurement act. Concerning it we first present the customary textbook version, which finds support in some parts of von Neumann's celebrated book.

3.2. Orthodox Theory of Measurement

The state of a physical system, characterized in general by a statistical operator ϱ, is capable of changing in two entirely different ways. One is the smooth temporal development which for pure cases is in accord with the Schrödinger equation. In general, the change is represented by the evolution operator $T(t_2, t_1)$ such that

$$\varrho_{t_2} = T(t_2, t_1)\, \varrho_{t_1}\, T(t_1, t_2). \tag{5}$$

This type of change involves a special form of causality, suitably named statistical causality, which allows the prediction (or retrodiction) of a probability distribution at a time t_2 when the distribution at time t_1 is given. It does not guarantee dynamical causality, the prediction of a single event at t_2 when a small set of dynamical variables at t_1 is known.

The second type of change in ϱ is said to occur when a measurement is made. It is abrupt, indeed it occurs for all practical purposes instantaneously and cannot be predicted; it has even been called an acausal jump. The orthodox theoretical statement of this change is known as the projection postulate[1], which asserts that after a measurement of A on a single system has yielded the eigenvalue a_n, that system is then in the corresponding eigenstate ψ_n. By equation (1), the measurement of A on an ensemble of systems with statistical operator ϱ leads to the expectation value $\bar{A} = \mathrm{Tr}(\varrho A)$. $\bar{A}$ may be expressed in this way:

$$\bar{A} = \sum_{n,m} (\psi_m, \varrho\, \psi_n)(\psi_n, A\, \psi_m) = \sum_{n} (\psi_m, \varrho\, \psi_m)\, a_m. \tag{6}$$

In the light of the projection postulate, equation (6) may be interpreted by supposing that after the measurement the fraction $(\psi_m, \varrho\,\psi_m)$ of the ensemble is in the state ψ_m, hence that the total ensemble is then characterized by the statistical operator

$$\hat{\varrho} = \sum_{m} (\psi_m, \varrho\,\psi_m)\, P_{\psi_m} = \sum_{m} P_{\psi_m}\, \varrho\, P_{\psi_m}. \tag{7}$$

Equation (7) is the measurement intervention transformation, the second kind of change in ϱ. Of particular interest is the effect of this transformation on a pure statistical operator $\varrho = P_\psi$:

$$\hat{\varrho} = \sum_{n} (\psi_n, P_\psi\, \psi_n)\, P_{\psi_n} = \sum_{n} |(\psi_n, \psi)|^2 P_{\psi_n} = \sum_{n} w_n P_{\psi_n}, \tag{8}$$

which means that an ensemble consisting initially of systems in the state ψ will be correctly described after measurement as a mixture of the ψ_n with respective weights

$$W_n = w_n \equiv |(\psi_n, \psi)|^2.$$

Having set forth the requisite background material, we now review the analysis of the problem of objectivity in quantum theory as given by the protagonists of the Copenhagen interpretation.

[1] Because of the mathematical form it takes in SCHRÖDINGER's wave mechanics, the transformation envisioned in the projection postulate is often called the "reduction of the wave packet."

3.3. Copenhagen Views on Objectivity

HEISENBERG's philosophic discourses on the Copenhagen interpretation of quantum theory [*14, 15*] note the presence in that theory of a suitably qualified subjectivity — qualified in the sense that any inference that quantum mechanical systems incorporate the mind of the observer is explicitly rejected. Though motivated by the broader Kantian subjectivity (cf. 1.5), Heisenberg bases his main argument that quantum theory is partly subjective upon the statistical elements of the theory; he is therefore primarily concerned with probabilistic subjectivity (cf. 1.5). In the sequel, the term *subjectivity* will refer only to this second meaning.

If subjectivity in quantum mechanics is to arise somehow in statistics, then classical Gibbsian statistics should similarly exemplify a subjective aspect of physics. Far from denying this assertion, Heisenberg builds upon it, observing that the probability function which characterizes the canonical ensemble assigns finite weights to all energies, even though the actual system under study has just one energy. Hence, "the canonical ensemble contains statements not only about the system itself but also about the observer's incomplete knowledge of the system" [*16*]. This two-fold referent for probabilistic statements is the central theme in HEISENBERG's arguments for subjectivity, both classical and quantum.

HEISENBERG acknowledges the difference in character between the probabilities of the classical canonical ensemble and the quantum pure case when he assigns complete objectivity to the latter. It is rather the manipulable probabilities that arise in quantum theoretical mixtures which Heisenberg regards as partly subjective because, in addition to objective statements about possible measurements, they contain "statements about our knowledge of the system, which of course are subjective in so far as they may be different for different observers" [*17*].

HEISENBERG's belief that quantum theory exhibits immutably some subjectivity seems to be founded on two propositions.

I. Reducible probabilities, being measures of knowledge which undergo sudden transformations whenever new facts become known, have a partially subjective nature. Hence mixture states are partly subjective.

II. The actual phenomena which are the objects of physical inquiry must necessarily be described as mixtures.

The background for proposition I has been discussed. Proposition II has its origin in certain features of the Copenhagen interpretation of BOHR [*18*] and HEISENBERG [*19*].

The latter insist that, inasmuch as the concepts of classical physics are employed in the description of actual experiments, the proper conception of a physical phenomenon must be so broadened as to embrace not only the behavior of the system S under study but that of the measuring instrument M as well. In the writings of both BOHR and HEISENBERG are found indications that it is the world view of classical physics which is correctly regarded as "real". This seems to be the root of BOHR's complementarity and of HEISENBERG's intriguing Aristotelian pronouncements that quantum states allude only to the "possible", whereas measurement induces a "transition from the possible to the actual" [*20*]. In particular, both advocate the principle that no contrivance may be regarded as a *measuring* apparatus unless its use involves an interaction of the compound system $S+M$ with the classical macroscopic world. Since it can be shown that after two originally independent systems interact, either system considered singly can only be left in a mixture state, it follows from these premises that a physical phenomenon must necessarily be described as a mixture, which is just proposition II.

Incidentally, the meaning of HEISENBERG's assertion that the state of a closed system is objective but not real [*21*], is now clear; for the denial of interaction with the classical world prohibits the "transition from the possible to the actual", which is tantamount to a denial of "reality" to the pure case.

The theme of proposition I, that subjectivity arises with the use of reducible probability notions in a theory, has been analyzed in section 2. What was said there in opposition to the subjective interpretation of probability is not affected by the restriction in proposition I to reducible probabilities. However, if the subjective interpretation of probability is accepted, then the position taken by HEISENBERG on the question of objectivity in quantum theory is logically unassailable. Proposition II, as explained above, is essentially a theorem in quantum mechanics; it is indifferent to the question of objectivity unless used in conjunction with proposition I.

At the 1957 Colston symposium, H. J. GROENEWOLD [*22*] argued that careful analysis fails to justify any subjective interpretation for quantum theory. Since his discussion is based upon the orthodox theory of measurement (cf. 3.2) accepted by the Copenhagen school, we briefly outline it, not only to acknowledge an interesting rebuttal to subjectivist views but also to focus more sharply on those features of quantum theory in which the Copenhagen subjectivity resides.

The GROENEWOLD argument consists of an orthodox quantum mechanical analysis of the following abstract experimental schema. Suppose that a given atomic system S is to be studied by means of a space ensemble of replicas of S. A typical sample of S is enclosed in one of those boxes cherished by physicists which shields its contents from the remainder of the universe. Included with S is a set of selected measuring instruments $M_1, M_2, \ldots$, which are triggered to perform measurements of observables $A_1, A_2, \ldots$ at successive times $t_1, t_2, \ldots$. Each such box is equipped with a sequence of shuttered windows behind which the measured eigenvalues appear. An observer gains cognizance of the measured results by opening the shutters. Whatever he learns by so doing may be incorporated into the statistical operator which describes S. Surely only these latter procedures could conceivably be construed as subjective in any sense. For it must be assumed that the capricious peeking of the experimenter can have no effect on the contents of the box. This very plausible assumption, however, is contradictory to BOHR'S stipulation that an essential characteristic of a measuring device is its irremovable interaction with the external world. In the light of GROENEWOLD'S schema, such a demand seems neither more nor less relevant that it would be in classical physics. The effect of observation on the pointer behind the window in this case is surely no more significant in a physical sense than, say, the effect of illumination on a falling body. Nonetheless, the interactions of S with $M_1, M_2, \ldots$ in the measurement act imply that S must be described as a mixture state. Thus, even with the present modification of basis for proposition II, the locus of HEISENBERG'S subjectivity — the mixture state — remains firmly embedded in the quantum mechanical description.

Now, looking into the windows on the boxes which house the ensemble will furnish information which permits selection of a subensemble, all the samples of which yielded, say, the same eigenvalue, a_k, of A upon measurement. This act of selection of subensembles is represented mathematically by a "reading" transformation which isolates whatever subjectivity, if any, there may be in this theory:

$$\hat{\varrho}_k = P_{\psi k}\, \varrho\, P_{\psi k} \tag{9}$$

where $\hat{\varrho}_k$ is the statistical operator of the k-th subensemble immediately after measurement[6]. Equation (9) is based on the principle of the reduction of the wave packet, as may be seen by considering $\varrho = P_\psi$:

$$\hat{\varrho}_k = P_{\psi k} P_\psi P_{\psi k} = |(\psi_k, \psi)|^2 P_{\psi k}$$

or when normalized, $\hat{\varrho}_k = P_{\psi k}$, an explicit mathematical statement of the projection postulate.

It is at least conceivable that transformations of this type, which do indeed seem to represent sudden changes in quantum statistical description engendered by increase in knowledge, might justify the claim of subjectivity in quantum theory. For the present outline it is enough to assume measurements at two times t_1, t_2, or A_1, A_2, respectively, where A_1, A_2 have simple discrete spectra. All that quantum mechanics can be expected to predict is summarized in $p_{a_1 a_2}$, which is the probability that the measured values of A_1 and A_2 were a_1 and a_2, respectively.

Suppose that between t_1 and t_2, the experimenter notes the relative frequency of the occurrence of a particular a_1 among the readings of A_1. Having thus selected an a_1'-subensemble, he "instantly, acausally, subjectively" alters ϱ by the "reading" transformation, then uses the new ϱ to predict the conditional probability that a_2' will be measured at t_2 in the a_1'-subensemble.

Let ϱ_{t_0} be the initial statistical operator. The procedure outlined above is expressed mathematically as follows:

(1) Time development to t_1:

$$\varrho_{t_1} = T(t_1, t_0)\, \varrho_{t_0}\, T(t_0, t_1)$$

(2) Measurement of A_1:

$$\hat{\varrho}_{t_1} = \sum_{a_1} P_{\psi a_1}\, \varrho_{t_1}\, P_{\psi a_1}.$$

(3) "Subjective act of selection":

$$\hat{\varrho}_{t_1 a_1'} = P_{\psi a_1'}\, \varrho_{t_1}\, P_{\psi a_1'}.^{1}$$

(4) Time development to t_2:

$$\varrho_{t_2} = T(t_2, t_1)\, \hat{\varrho}_{t_1 a_1'}\, T(t_1, t_2).$$

The desired (relative) probability is now given by

(5) $p_{a_1' a_2'} = \mathrm{Tr}\,(\varrho_{t_2} P_{\psi a_2'})$:

$$= \mathrm{Tr}\,[T(t_2, t_1)\, P_{\psi a_1'}\, T(t_1, t_0)\, \varrho_{t_0}\, T(t_0, t_1)\, P_{\psi a_1'}\, T(t_1, t_2)\, P_{\psi a_2'}].$$

To what extent that reading and "subjective" adjustment of ϱ affect the quantum theoretical prediction can now be assayed. This is accomplished by calculating the probability $p_{a_1' a_2'}$ *without* benefit of the reading of a_1 in the course of the experiment. When this is done, there comes the revelation that the result is the same whether the experimenter looks or not. But step (3) can no longer be called "subjective". Thus GROENEWOLD concludes that allegations of subjectivity in quantum theory arising from notions about discontinuous changes in the observer's knowledge are essentially verbal.

3.4. Critique of the Projection Postulate

The projection postulate is fashioned after classical science, where one knows that after a system has been found (measured) to be in a certain state it will be there if one looks again immediately afterward.

[1] This expression is not normalized.

In quantum mechanics, the situation is not quite so simple. In the first place its systems are so delicate that a measurement may alter their states *un*predictably so that even if a value a_i emerges one cannot be sure that it is left in the state ψ_i — indeed a measurement may destroy the system altogether. Secondly, there emerges a more subtle, theoretical difficulty. If the projection postulate is correct, a single measurement, in yielding a determinate ψ_i, would suffice to create knowledge of an entire probability distribution to which ψ_i is related by equation (2). Thus, quantum mechanics is hardly a normal stochastical theory where, as a general rule, a single observation cannot determine a complete distribution.

At this juncture our account is reminiscent of one of the versions of the probability concept, namely the subjective one. According to it, ψ is a measure of *knowledge* of an experimenter before he performs a measurement. When the measurement has been made, knowledge of the outcome is definite and w has jumped to 1 for the property i actually observed, the state from ψ to ψ_i. Hence the projection postulate in the form stated above conveys the subjective meaning of probabilities. Furthermore, barring theories of hidden variables [*23*], states in the form of ψ-vectors are the last instance of appeal in atomic theories, and if they are truly measures of personal knowledge which fluctuates with incidental evidence quantum mechanics must indeed be regarded as a subjective description of man's experience in the sense of all the versions of objectivity presented in section 1.

There is, however, a more cautious variant of the projection postulate, expressed mathematically by equation (9). It acknowledges on the whole the abrupt changes which that axiom envisions but refuses to consider them acausal. A measurement, it affirms, performs a *selection* of systems from an ensemble, thus generating a subensemble containing a smaller number of systems but all in the same measured state. Accordingly, ψ refers to the original ensemble, ψ_i to the subensemble.

We note that this understanding of the projection postulate restores objectivity. The systems found in the post-measurement ensemble were already there originally; if the subensemble to which ψ_i belongs were included among the totality of systems or, in case there is but one system present, if the observations yielding a_i were included among all observations made, ψ would be the same before and after measurement. The "jump" from ψ to ψ_i does in fact have reference to knowledge, but not merely to personal, subjective knowledge. An objective circumstance, the same for all, corresponds to the change of knowledge which is likewise the same for all observers. The selection is an objective procedure in accordance with 1.2, 1.3, and 1.4.

Thus arises the question: which of the two implications of the projection postulate is correct or, if neither, which is nearer the truth? The "truth" in this instance is furnished by a more elaborate theory of measurement which, perhaps strangely, can, like the orthodox theory of 3.2, also be drawn from VON NEUMANN's book. This theory assumes only the generally accepted postulates of quantum mechanics. Since it is developed in several places [*24*, *25*], we confine our discussion to its major results.

Suppose that a pure case, ψ, of a given system S is present and that a measurement of A is made. When the interaction is followed through in mathematical detail, it is seen that ψ becomes entangled with the state of the measuring apparatus M so that ψ_S and ψ_M convert themselves into a new state ψ_{SM} defined in the combined Hilbert space of S and M. This conversion takes place in strict obedience to the Schrödinger equation and there is no suggestion of an acausal jump. But if now the statistical operator is computed for the state ψ_{SM} in the Hilbert space of S *alone*, it turns out that this operator is not a projection (pure case), but is given by equation (8), a mixture. Indeed the reducible probability with which state ψ_n appears in the mixture has exactly the value given by equation (3). The measurement intervention transformation of the orthodox theory of measurement is therefore *derived* without using the projection postulate at all.

It is already apparent that the first interpretation of the projection postulate is untenable. Measurement changes the state from ψ to a mixture which cannot be written as a single ψ_i. And that is all it does; moreover, it accomplishes this in (stochastically) causal fashion, without violating the Schrödinger equation. But of even greater interest is the fact that the probabilities which appear in the resulting mixture are *reducible*, i.e., manipulable in physical ways. The process whereby one of them is converted to 1 while all other probabilities take on the value 0 is called a selection. Hence the second interpretation of the projection postulate is not far off the mark. While a measurement, in the widest sense, *need* not effect a selection of systems, all in the measured state, it *can* often be coupled with a procedure that achieves this end. We may therefore accept the second explanation as *permitted* by the formalism of quantum mechanics even though we might wish to grant that *there are measurements which are not selective.*

What matters here, however, is this conclusion. The one basic proposition of quantum mechanics which threatens objectivity in at least one of the forms we have discussed is the projection postulate, and it holds this threat only in its first interpretation. That interpretation is erroneous when analyzed fully. Hence the threat is removed. Quantum states, even though they correspond to probabilities, are objective provided the theories of 1.2, 1.3, or 1.4 are accepted.

REFERENCES

[1] ROSENFELD, L.: Strife about complementarity. Sci. Progr. **41**, No. 163, 406 (1953).
[2] BUNGE, M.: Strife about complementarity. Brit. J. Philosophy Sci. **6**, No. 21, 9 (1955).
[3] BORN, MAX: Physics in my Generation, p. 158—163. London: Pergamon Press 1956.
[4] MARGENAU, H.: The Nature of Physical Reality. New York: McGraw-Hill Book Co. 1950.
[5] HEISENBERG, W.: Physics and Philosophy, p. 58. New York: Harper & Row 1958.
[6] BORN, MAX: Symbol and Reality, Appendix 3 in his Natural Philosophy of Cause and Change. New York: Dover Press 1964.
[7] EDDINGTON, A. S.: The Philosophy of Physical Science. Cambridge University Press 1939.
[8] — Space, Time, and Gravitation. Cambridge 1920.
[9] Sir J. JEANS: Physics and Philosophy, p. 178. London: Macmillan & Co. 1943.
[10] — Op. cit., p. 170.
[11] POPPER, K. R.: In: Observation and Interpretation in the Philosophy of Physics, p. 65—70. New York: Dover Press 1957.
[12] MARGENAU, H.: Op. cit., esp. Ch. 13.
[13] NEUMANN, J. VON: Mathematische Grundlagen der Quantenmechanik. Berlin: Springer 1932.
[14] HEISENBERG, W.: In: NIELS BOHR and the Development of Physics [W. PAULI (ed.)], p. 12—29. New York: McGraw-Hill Book Co. 1955.
[15] — Physics and Philosophy. New York: Harper & Row 1958.
[16] — Physics and Philosophy, p. 138.
[17] — Physics and Philosophy, p. 53.
[18] BOHR, NIELS: ALBERT EINSTEIN: Philosopher-scientist, p. 199. New York: Harpers 1949.
[19] HEISENBERG: Op. cit.
[20] — Physics and Philosophy, p. 54f. See also the theory of latent observables discussed in ref. [4].
[21] — In: NIELS BOHR and the development of physics, p. 27. New York: McGraw-Hill Book Co. 1955.
[22] GROENEWOLD, H. J.: In: Observation and Interpretation in the Philosophy of Physics (S. KORNER, ed.), p. 197. New York: Dover Press 1957.
[23] See MARGENAU, H.: Advantages and Disadvantages of Various Interpretations of Quantum Mechanics. Physics Today **7**, 6 (1954).
[24] LONDON, F., and E. BAUER: La Theorie de l'Observation en Mécanique Quantique. Paris: Hermann 1939.
[25] MARGENAU, H.: Measurements and quantum states: part I. Phil. Sci. **30**, 1 (1963); part II. Phil. Sci. **30**, 138 (1963).

Chapter 11

Scope and Limits of Axiomatics

Paul Bernays

Eidgenössische Technische Hochschule
Zürich, Switzerland

When today one speaks on axiomatics to a public familiar with mathematics, there often seems to be not so much a need for recommending axiomatics as to warn against an overestimation of it.

In fact, there are today mathematicians for whom science begins only with axiomatics, and there are also mathematically minded philosophers, especially in Carnap's school, who regard axiomatization as belonging to the construction of scientific languages. Thus science is then regarded as being essentially deductive, whereas in fact only some sciences, and even so in an advanced stage, proceed mainly by derivation.

There are indeed empirical sciences where certain kinds of objects, for instance plants, are investigated as to their various forms and behaviors, as to the ways of their occurring as well as to the structure of their development. Here one obviously has first to care for acquiring a sufficient empirical material, and a premature axiomatization might divert one from a sufficient exploration of what is to be found in nature, or at least it may restrict this exploration.

On the other hand we have to be conscious that collecting and presenting material has a scientific value only if this material gives rise to suitable conceptions, classifications, generalizations, and assumptions. A mere material, without any theoretical aspect joined to it, cannot even be kept in mind, at least if it has a considerable extent. It is true that tables of collected material can prove to be highly valuable even in a way not originally taken into account; but certainly this takes place only if the material is brought in connection with directive ideas, i.e. with some conceptuality.

Now, whenever some conceptuality develops, when the question is one of putting results together, or of evaluating heuristically a problem situation, or also of subjecting assumptions to empirical tests, or even of investigating their inner consistency — in all such cases axiomatic thinking can be very fruitful. In fact it is then a matter of making clear what exactly the results are yielding, or what properly the problem situation is, or in what the assumptions strictly consist. Such a strengthened consciousness is valuable whenever the danger exists that we may be deceived by vague terminology, by ambiguous expressions, by premature rationalizations, or by taking views for granted which in fact include assumptions. Thus, the distinction between inertial mass and gravitational mass makes it clear that their equality is a physical law — something which might be overlooked by speaking of mass as the quantity of matter. Various instances of such a kind were especially considered by ERNST MACH.

Another sort of occurrence of axiomatics in the development of science is the presentation of a new directive idea in axiomatic form. A famous instance of this is the presentation of thermodynamics by CLAUSIUS, starting from the two main laws; and, of course, the classical instance of NEWTON's presentation of his mechanics cannot be omitted in this connection.

For both the said methods of applying axiomatics one can find also examples in biology, in particular the theory of descent and genetics, as well as in theoretical economics.

All the axiomatic systems considered so far have the common trait that they sharpen the statement of a body of assumptions, either for a whole theory or for a problem situation, and of being embedded in the conceptuality of the theory in question; eventually, then, this conceptuality might be somewhat sharpened by it. We might perhaps call them "material" or "pertinent" axiomatics.

In connection with pertinent axiomatics it may be mentioned that some important enterprises in theoretical science, in particular in physics, have an axiomatic aspect — namely all those where some discipline comes to be incorporated in another discipline, or where two disciplines are contracted into one theory. Well known instances are the incorporation of thermodynamics into mechanics by the methods of probability theory and statistics, and the embedding of optics in the theory of electromagnetic waves, as well as the joining of physical geometry with gravitation theory in general relativity theory. Of course we do not have here simply axiomatics, since the said reductions were partly motivated and also had to be justified by experimental results.

Nevertheless such reductions can also be considered under the aspect of that kind of axiomatic investigation which HILBERT called "the deepening of the foundations" of a theory and of which he gave many instances in his article "Axiomatisches Denken".

Examples of such a deepening are: the proof of the solvability of every algebraic equation, with one unknown, in the domain of complex numbers — though the modern algebraists do not admit this theorem as an algebraic one, at least in its usual presentation —, and the reduction of the theory of real numbers to that of natural numbers with the addition of the concept of number set or else that of a denumerable sequence.

The methods used in the deepening of foundations gave rise to a kind of axiomatics different from pertinent axiomatics and which one might call definitory or descriptive axiomatics, to which belong abstract disciplines developing the consequences of a structural concept described by axioms. Examples of such concepts are those of group, lattice, and field.

There is also the possibility of combining such concepts. Thus the metric continuum can be characterized as an ordered field whose order satisfies the Dedekind condition of absence of gaps.

A remarkable circumstance in descriptive axiomatics is that there is a multiplicity of equivalent characterizations. Thus lattice theory can be axiomatized just as well with the concept of equality or with the relation "sub" as a fundamental predicate. In group theory we can postulate the existence of a right-side unit element and a right-side inverse; or else we may postulate the solvability of each of the equations $ax=b$ and $ya=b$. Hence axiomatization contains elements of arbitrariness. The newer axiom systems for geometry, in particular those of HILBERT and of VEBLEN, are also descriptive, in contrast with the axiom system of PASCH, which is intended as a pertinent one. In the geometrical axiom systems a great freedom in the choice of kinds of individuals and the fundamental relations is to be found.

You may wonder why I have not yet mentioned the most famous instance of axiomatics: Euclid's *Elements*. But from the point of view of method this instance is not simple.

According to the traditional view the Euclidean system is to be regarded as a pertinent axiomatics, yet not starting from empirical facts but rather from intuitively evident facts. You know that there has been and there still is much debating about geometrical self-evidence. Yet however this question might be decided, Euclid's enterprise cannot be understood as attempting to secure a high degree of intuitive evidence. The interest lies here certainly in the possibilities of deductive reduction.

The way of taking the geometric axioms as purely hypothetical, as it is predominant in to-day's mathematics, is, at least, strongly prepared in Euclid's Elements. Only the axioms on quantities (the κοιναὶ ἔνοιαι) have here a separate status. (It is taken as self-evident, and it is not even axiomatically formulated, that the lengths of segments, the angles, areas and volumes are quantities satisfying those axioms.)

It was also a merit of Euclid's axiomatization that attention was called to the special role of the parallel axiom and thereby mathematicians became later on induced to discover non-Euclidean geometries.

In contemporary mathematics geometry is treated in many different disciplines: analytic geometry — in particular vector geometry —, differential geometry, projective geometry, algebraic geometry, and topology. FELIX KLEIN distinguished the kinds of geometry by the group of transformations which preserve the properties studied by the discipline.

Under the influence of symbolic logic one has passed from axiomatization to formalization. In the latter,

(1) the possible forms of statements in a theory are delimited a priori, and

(2) the logical inferences are subjected to explicit rules.

It is possible to formalize in this way the existing proofs in number theory, infinitesimal analysis, set theory, geometry, etc.

Nevertheless formalized axiomatics cannot fully replace descriptive axiomatics. The reason is that the sharpening of the concepts of predicate, function, sequence, and set, which is brought about by formalization, includes a restriction of these concepts to the effect that interpretations of the formal system are possible which are excluded by the corresponding original descriptive axiomatics.

This deficiency consists for the formalization of any axiom system wherein one of the said concepts is used, as in the induction axiom of number theory or in the continuity axiom of analysis.

It appears in a twofold way:

a) in a "syntactic" way, by the impossibility of formally proving certain number-theoretic theorems, which hold true upon the consistency of the formal axiom system, — as it was stated by GÖDEL.

b) in a "semantic" way, by the existence of "non-standard models", i.e. models differing from the structure to be described by the axiom system, — as stated first by SKOLEM.

So there seems to be a limit to strict axiomatization in the sense that we either have to admit a certain degree of imprecision, or to be unable adequately to characterize what we mean, for instance, by well ordering, continuity, and number series.

Biographical Notes on the Contributors

PETER GABRIEL BERGMANN. Ph. D. (Prague). Professor of Physics, Syracuse University. Taught at Princeton University and Yeshiva. Former coworker of EINSTEIN and a leading authority in general relativity. Author of *Introduction to the Theory of Relativity, Basic Theories of Physics, The General Theory of Relativity* (in the *Encyclopedia of Physics*) and many articles.

PAUL BERNAYS, Ph. D. (Göttingen). Professor Emeritus of Mathematics, Eidgenössische Technische Hochschule, Zürich. Taught at Zürich, Göttingen, and University of Pennsylvania. Has done decisive work in logic, set theory, foundations of mathematics, and theory of knowledge. Co-author, with D. HILBERT, of *Grundlagen der Mathematik* (2 vols.) and, with A. A. FRAENKEL, of *Axiomatic Set Theory*, as well as author of numerous articles. Past President, Académie Internationale de Philosophie des Sciences.

MARIO BUNGE. Ph. D. (La Plata). Distinguished Visiting Professor of Philosophy and Physics, University of Delaware. Taught at Buenos Aires, La Plata, University of Pennsylvania, Freiburg, etc. Author of *Cinemática del electrón relativista, Causality, Metascientific Queries, Etica y ciencia, Intuition and Science, The Myth of Simplicity, Scientific Research, Foundations of Physics,* and many papers in theoretical physics and philosophy. Member of the Académie Internationale de Philosophie des Sciences.

HAROLD GRAD, Ph. D. (New York University). Professor of Mathematics and Director of the Magneto-Fluid Dynamics Division of the Courant Institute of Mathematical Sciences, New York University. Has performed extensive work on plasma physics, statistical mechanics and thermodynamics.

EDWIN T. JAYNES. Ph. D. (Princeton). Professor of Physics, Washington University. Has worked in statistical mechanics, mathematical statistics and microwave development. Author of *An Electronic Theory of Ferroelectricity*.

PETER HAVAS. Ph. D. (Columbia). Professor of Physics at Temple University. Taught at Lehigh. Member of the Institute of Advanced Study at Princeton and resident research associate at Argonne. Engaged in distinguished research in general relativity, electromagnetic theory, and elementary particle theory. Author of numerous papers.

HENRY MARGENAU. Ph. D. (Yale). Eugene Higgins Professor of Physics and Natural Philosophy at Yale. Distinguished research in quantum mechanics, theoretical spectroscopy and plasma physics. A leading authority on the philosophy of physics. Author of *The Nature of Physical Reality* and two hundred other scientific and philosophic publications. Past President, Philosophy of Science Association. Member, Académie Internationale de Philosophie des Sciences.

WALTER M. NOLL. Diplom-Ingenieur (Berlin) and Ph. D. (Indiana). Professor of Mathematics, Carnegie Institute of Technology. Taught at the University of Southern California. Accomplished distinguished research in continuum mechanics and in the foundations of classical mechanics. Co-author with C. TRUESDELL of *The Non-Linear Field Theories of Mechanics* (in the *Encyclopedia of Physics*) and author of numerous papers.

E. J. POST. Ph. D. (Amsterdam). Investigator, Air Force Cambridge Research Laboratories. Worked for the Netherlands Telecommunications Laboratory and the Bell Telephone Laboratories. Engaged in research on piezo-electric materials and electromagnetic theory. Author of *The Formal Structure of Electromagnetics*.

RALPH SCHILLER. Ph. D. (Syracuse). Professor of Physics, Stevens Institute of Technology. Taught at Syracuse and the University of São Paulo. Well known for his work on general relativity, classical models of quantum mechanics, and spinning particles.

CLIFFORD A. TRUESDELL, III. Ph. D. (Princeton). Professor of Rational Mechanics, Johns Hopkins University. Worked at the U.S. Naval Research Laboratory and taught at Maryland and Indiana. Authority on continuum mechanics and specialist in the foundations and the history of mechanics. Co-author of *The Classical Field Theories* and *The Non-Linear Field Theories of Mechanics* (both in the *Encyclopedia of Physics*) and author of *Six Lectures on Modern Natural Philosophy* and numerous papers. Founder and editor of the *Archive for Rational Mechanics and Analysis* and of the *Archive for History of Exact Science*.

Druck der Universitätsdruckerei H. Stürtz AG., Würzburg